# HOW TO BEAT THE ODDS

## BE DRIVEN!

EDDIE MADDOX, PH.D.

Copyright© 2020 Eddie Maddox

All rights reserved. No part of this book may be reproduced or transmitted in any form, without the written permission of the author, except for brief quotations for review purposes.

Published by ELOHAI International Publishing & Media:
P.O. Box 64402
Virginia Beach, VA 23467
elohaipublishing.com

ISBN: 978-1-7348778-6-1

Printed in the United States of America

# DEDICATION

This book is dedicated to my family. My intelligent and beautiful wife, Candace, has been beside me through tough and easy times. We started out as a couple that was focused on building a successful family using the founding principles of this country. Our focus is to give liberty and justice to all. She motivated me to write this book because she saw where I started in life and she has encouraged me to strive for the very best. This woman is extremely precious and is clearly my better half. Our two children, Brian and Kristin, inspired me to continue my educational goals. I saw them working on class work, and it motivated me to reach for higher knowledge. My son also inspired me through his dedication and work ethics. He is an inspiration for any young child and adult, academically, athletically, and professionally. My daughter has motivated me by her hard work and artistic skills. She is an excellent role model for others to emulate. She too has inspired me through her work, academically and professionally.

I want to personally thank my wife and children for inspiring me to challenge the status quo and reach for higher learning. I have learned from all three of them that education can change a person's ability to solve problems and help make this great world a better place for all living things. I dedicate this book to them.

# ACKNOWLEDGMENTS

My parents are the first people I would like to acknowledge. They both encouraged me to be the best at whatever I set out to accomplish. They taught me the importance of school. As a young student, I was fortunate to have wonderful teachers in Head Start, elementary, and middle school. My good fortune continued through high school and college. I would like to acknowledge all my teachers because each one helped create and mold me into an individual who always strives for more knowledge. My Head Start teacher, Mrs. Elsie Thompson, created a great foundation in me. She taught me the importance of being a role model student and the necessity of changing society for the best by learning new things and being involved. I want to give two of my elementary school teachers special thanks. Mrs. West and Mrs. Pat Evans were excellent teachers in the early years of my education experience. They taught me the key skills to being a successful student. In middle school, I met a physical education teacher who gave me purpose and meaning.

Coach Shabazz taught me the value of physical fitness and the importance of taking care of the mind and body. He gave me the courage to always challenge myself to be the very best. Also, in middle school Mrs. Williams taught me the importance of utilizing my talents in mathematics. I consider her not only a math teacher but a counselor who made sure I enrolled in the correct classes as I moved on to high school. Mrs. Williams' concern was the greatest gift any teacher could have given me. I also had a terrific teacher in high school. She was my math teacher, Mrs. Emma Doughty. She would not allow me to doubt myself. Thanks to all these outstanding teachers, I have been able to achieve my goals.

I also want to acknowledge Jonathan Waller. He is a great friend and confidant. It is amazing how God knows how to align you with the right people. God brought Jonathan into my life when I was engaged in a project at work. He and I knew at our first meeting that we were meant to be friends. Jonathan has a strong faith and through him I learned the importance of fasting and prayer. Terence Hassan is another friend I want to acknowledge and thank for creating the spark inside me for this book. He told me one day that I had an awesome story to tell of how I beat the odds. His words are responsible for me sharing the details in this book with others so that they can use this information to learn how to beat the odds themselves.

To all those who have participated in this book, I acknowledge each of you for your dedication in making a contribution. Your service and input are very significant for helping others begin the journey towards success.

# PRAISE FOR *How to Beat the Odds*

"This book is about championing the cause of eradication of debt and overcoming the fears of obtaining financial freedoms to pursue happiness. Dr. Maddox is best suited to write this book as he never has forgotten his humble beginnings even through his great success and accomplishments. I've seen him cut his entire yard with a push mower while having a riding mower in his garage. I've also seen him drive a Ford ranger pickup truck while having a Mercedes in his garage. I have never witnessed him talking down to or belittling anyone, his deeds or words have always been exceptional. Eddie's dedication to tell this story is a testament to the man that he is, and I hope that his book will encourage and motivate others to be all that they can be by gaining their financial independence. It is my hope that you will read it and learn from Eddie's vast knowledge and experience."

- *John Gambrell*, **Army Sergeant - Cousin and Friend**


"Aspiring leaders would be remiss to fail to embrace the pragmatism, optimism, and unequivocal drive of Dr. Eddie Maddox. Dr. Maddox, scarred by experience, has forged himself, through attitude and aptitude, into a successful leader, entrepreneur, mentor, and educator. Dr. Maddox's approach challenges the Peter Principle, with conviction, and removes the concepts of incompetence, or inability, from the vernacular. I am sincerely excited for the cumulative opportunity to study, and ultimately emulate, the behaviors of a mentor, for which I am eternally grateful."

- *Chris Burdette*, **Director of Operations and Former Mentee**

"As a teacher, my career began in Alabama in 1958 and spanned thirty plus years during which I was honored to teach thousands of children in grades four through twelve. In every teaching career there are a few students that "flip the script" and inspire the teacher in a unique way, and Eddie Maddox was that student for me. While Eddie has always maintained that I influenced him, I maintain that the opposite was true. Eddie was an "old soul" even as a fourth grader. He had a sense of purpose and character that was extraordinary for one so young. However, despite his natural gifts, he also displayed a quiet determination for excellence and a receptive spirit for learning. I can only acknowledge that I recognized his giftedness (but I couldn't imagine how anyone could not have noticed). While I certainly wish that I could claim credit for Eddie's many achievements, both personal and professional, I must conclude that he was destined and self-determined for success –and he has most assuredly earned it. I can only be honored to have been a part of his journey."

- *Pat Evans*, **Fourth Grade Teacher Lifetime Friend**

"This is the story behind the accomplishments and success that Eddie has experienced. Eddie has a level of commitment that creates a mindset to not give up and work until he succeeds. Sometimes success means you deviate from your original plan, but it still gets you to the next level. This commitment was critical in him accomplishing goals even when circumstances forced him to take a different route. The most important thing that Eddie has is his faith in Jesus Christ to carry him through anything that this world can offer. This faith provides a foundation that allows his pride and commitment to fuel his drive to accomplish many things. I have enjoyed working and becoming friends with Eddie over the past four years as we have similar life story journeys."

- *Marty Hallman*, **Plant Manager**

"On this journey that we call life, there are people that come into your life for a reason, a season, or a lifetime. Eddie is an individual that will leave a lasting impact under any of these circumstances. This is the story of a young man that went from being an at-risk kid in the Head Start program to a calculated, risk taking individual that has been successful in sports, education and business. This is the story of an individual that defied the odds and has been able to rise from poverty as a child to being in the top one percent of wage earners in America. In traveling this journey with Eddie, you will discover certain Godly principles that can be applied regardless of background, religion, race or ethnicity. Eddie has leveraged his talents and abilities, helped others along the way and operates with the highest level of integrity. Take this journey and discover the principles that can help anyone to beat the odds."

- *Dr. Jonathan A. Waller*, **Plant Manager**

# TABLE OF CONTENTS

A Letter to the Reader.................................................. 12

Foreword.................................................................. 17

Introduction: The Twenty Success Factors................... 24

Chapter 1: Childhood Roots........................................ 31

Chapter 2: Maintain Your Focus Despite Naysayers.... 37

Chapter 3: Make Education Your Foundation for Success....... 41

Chapter 4: Identify your Leadership Qualities............. 49

Chapter 5: How to Overcome Obstacles...................... 67

Chapter 6: Achieving Financial Success...................... 77

Chapter 7: Honor Important Relationships.................. 89

Chapter 8: Make God Your Foundation....................... 101

Chapter 9: Learn from Coaches and Teachers............. 121

Chapter 10: Set Clear Goals and Measure Progress..... 129

Chapter 11: Drive to Achieve...................................... 137

About the Author...................................................... 165

## HOW TO BEAT THE ODDS-BE DRIVEN!

# A LETTER *To The Reader*

Dear Reader,

My name is Brian Maddox, and I am the son of Eddie Maddox, Ph.D., the author of this book. Before you dive deep into the pages that follow, I want to encourage you to commit yourself to finishing this book. The lessons that you now hold in your hands can and will change the trajectory of your life.

My father offers wisdom to all people that he meets and works with each day. This book displays wisdom for those who want to work towards success. He instilled all the values in this book into my sister Kristin and me. Following the words of wisdom that he offers in this book has allowed me to play college football at the University of South Carolina and graduate in three and a half years.

He taught me to make God my foundation. He was and still is a strong family man today. He spent quality time with me and my sister as we were growing up and he taught us life lessons that have prepared us for our journey towards success. He also taught us to build a strong circle of friends. You need others to help you get

## A LETTER TO THE READER

through the journey towards success and he taught us to help others on their journey towards success. He stressed the importance of an education. He showed us that an education, more than anything, can help determine the level of success that we will achieve. He exposed us to how to manage money at a young age. He put emphasis on the following areas: never live above your means, be patient, and save and invest to achieve even more. He told us at an early age that we would run into obstacles in whatever we did, but he taught us to never give up because we could make it over those obstacles.

My dad taught us not to hold pity parties when things were not going our way. He taught us to determine how to overcome those challenges instead. He also stressed the importance of being coachable and learning from others. Once I graduated from the University of South Carolina, the wisdom he shares in his book helped put me on a path of success in management in the logistics industry. I recommend that you read this book so that you can learn valuable information that can help you on the path to success. Lastly, I believe that others should take advice from this book because of my father's hardworking mentality and his drive to always make things better (Continuous Measurable Improvement). In all his success, my dad is committed to always giving back to the community through love, coaching, mentoring and collaborating with others.

# HOW TO BEAT THE ODDS-BE DRIVEN!

**These are three of the biggest takeaways that I've learned from my father and that you will glean from the pages of this book:**

1. Hard-working mentality: He always starts with a plan on how he will solve whatever the task might be. He will take action and never give up when it gets challenging. Nothing screams "hard work" more than pursuing higher education, especially the daunting task of earning a Ph.D.! Finally, he always goes back to review the results.

2. Drive to always make things better (Continuous Measurable Improvement): He has always taught my sister and me that anything can be improved. From personal experience, when I would have my best games in high school or college football, we would always analyze what I could have done differently. When you look for ways to improve and never get to the point that you think you have arrived, then your success level in life will be astounding. Thanks to him, I now use this in the business world and have found a tremendous amount of success!

3. Giving back to the community: I have always known him to care about the community. This has been evident in various different ways: coaching Little League sports in the town in which he grew up, serving with the Chamber of Commerce, serving on boards, speaking to groups about key ways to be successful, and just being an overall leader in the community!

# A LETTER TO THE READER

We love him and will continue to use all of these key factors in this book in our everyday lives! This book can be a game changer for you as you take the journey toward success.

Sincerely,

Brian Maddox

*Be involved with things that matter to this world.*

*- Eddie Maddox, Ph.D.*

# FOREWORD

If you were dropped off by helicopter into the unfriendly wilderness, with no training, and all you had was a pocket-knife, a small backpack, and a piece of flint, would you survive? For most of us the answer would be NO! Of those who would survive, how many would thrive? Eddie is one of those very few who has not only survived, but thrived in the wilderness, and I would put my money on him.

Eddie's life start can be best described using the first two lines of Stevie Wonder's song "Living for the City." It goes like this: "A boy is born in hard time Mississippi. Surrounded by four walls that ain't so pretty." Eddie's book starts here except in South Carolina, not Mississippi, and takes you on his life's journey. Vegas would probably bet against Eddie being successful. The oddsmaker would probably say the odds are five to one that he would not make it to the twelfth grade, ten to one that he would not graduate from college, twenty to one that he would not graduate from a four-year university with a degree in manufacturing engineering. So if you were a betting individual you would probably bet with Vegas and lose your money.

## HOW TO BEAT THE ODDS-BE DRIVEN!

Having grown up in the same town with Eddie, I watched him play football starting at the age of ten. When I returned home from college I volunteered and worked with the high school track team in which Eddie was a sprinter. Years later Eddie and I formed an Amateur Athletic Union (AAU) team and coached a summer track team together. Our children played on the team, attended the same public schools, and our wives and families shared common interests. All of the above have allowed me to observe this remarkable man's journey through life. This observation advantage has prepared me to write the Foreword for his book, and for this, I am honored.

We all possess special gifts, tools, and talents. We are born with them. Some of us never develop them, while others excel in certain areas. When you see someone who excels in what they do, it is usually because they have perfected their gift or talent. They become so good at what they do that they make it look easy (i.e. Steph Curry, NBA professional). They tend to dominate their competition at times (i.e. Simone Biles, Olympic gymnast). It is so obvious that they are a better athlete, pianist, or teacher. Eddie showed at an early age that he was born to be a football running back. I had heard about Eddie's running style and speed at the barber shop in our home town, so I went to a game. After the first quarter of his game, I said

# FOREWORD

to myself, "This kid is the real deal." He had all of the gifts and tools needed to be a successful athlete. He had the speed, lateral quickness, great acceleration, he ran with a low center of gravity, and most importantly, he was smart and ran with an attitude. With all of that talent the big question was, "Will he make it out of the wilderness?" As this book reveals, he and his siblings were surrounded by four walls that were not pretty. Eddie, like so many others, would have obstacles to get around and jump over. I am proud to say, years later, that he made it out of the wilderness and I've been here to witness his success.

All of us possess at least one or two characteristics that are dominant in our personalities. For Eddie Maddox, I see determination. That is one reason why I enjoy engaging in conversations with him when I see him around town and at events. When I spend time with Eddie, his positive energy leaves me feeling fired up. Determination was and still is the key vehicle that has gotten Eddie to where he is today. A physically challenged athlete, Bill Demby, once said, "There aren't really enough crutches in the world for all the lame excuses." Eddie does not entertain excuses as a reason for not achieving. His determination prohibits that.

# HOW TO BEAT THE ODDS-BE DRIVEN!

The second strong characteristic that I see in him is drive. We were on the phone one day and Eddie told me that he was working on his Ph.D. I was speechless for a moment as I was thinking, "Here is a guy who has already accomplished so much." He finished his collegiate career as an Academic All-American football player, he graduated with honors, and he moved through the ranks to become the quality and operations manager for a major manufacturer, just to name a few of his accolades. And now he's working on his Ph.D.! Eddie is fifty-plus-years-old and still has so much drive.

Eddie grew up to be a "family first" provider. One could see that this was monumentally important to him. Eddie and his wife, Candace, were high school sweethearts, so their relationship was well established by the time they married and had children. They seemed to have been of one accord when it came to values and how to raise their children. Eddie and Candace were very active and supportive parents when it came to their children. You would see them attending their children's school activities such as open houses or parent teacher conferences. Our sons participated in the same science fair one year. You would have thought that those were our projects instead of our sons'. We were both proud of our kids' projects, and were "in it to win it." A main focus for Eddie and Candace was to equip their

# FOREWORD

kids with the tools they needed to be successful. Eddie served as president of T.L. Hanna High School Booster Club during the years his son Brian played high school football. Brian ultimately was awarded a full scholarship to play for the University of South Carolina. When the stakes are high, you have to be there for your kids. Eddie made sure to do that by staying close and being political. Kristin, their daughter, a Clemson University graduate, and Brian have very promising and successful careers. Their parents' commitment to providing a stable home environment, instilling coping skills, and holding them accountable for their grades and behavior have paid off.

I urge you to share this book with young people. It is a great example of how a small town guy beats the odds life dealt him. Use it as an "example manual" for young people. You have survival manuals and manuals for first aid. They are used as instructional guides, refresher courses, and training. Use this book to do the same. I strongly suggest that parents, mentors, teachers, coaches and youth ministries use this to teach key concepts that the author embraces. Young athletes, male and female, will relate to Eddie's attitude and his approach to his academics and sports as a student.

## HOW TO BEAT THE ODDS-BE DRIVEN!

You can also use this book if you are "not so young." If your tires have been running for a while and the tread is getting a little thin, it is not too late to change tires. Most of us perform way below our capabilities mainly due to a lack of confidence and/or motivation. Some of your talents may have become old, dusty or rusty, but there are some things in these chapters that can help you hit the reset button in your life. Use the motivation and "aha moments" that this book gives you to make small changes. It's the small changes that can make a big difference.

Having already co-authored a book for the manufacturing industry, this book about Eddie's life is way overdue. Just to be clear with you, Eddie is a very modest person, so you will learn more about him from reading this book than you will spending a full season watching football games with him. He does not talk about himself much. He does like talking about "life's lessons." This book captures what's in Eddie's head, heart, and soul.

- Terence Hassan, **State Farm Agency Owner**

*Be the conductor of your life.
You are in control of your destiny.
Do not allow others to control it for you.*

*- Eddie Maddox, Ph.D.*

# INTRODUCTION:
## *The Twenty Success Factors*

This is a book of wisdom and important lessons that I've learned on my journey toward success. I am an African-American male who lived in poverty as a child, yet I climbed the success ladder to become amongst the top one percent of wage earners in America. I defied the odds to become a successful, family-centered man who teaches others to be successful regardless of their background, religion, race, or ethnicity. I started out in the Head Start program and worked to achieve my Ph.D. I had parents who believed in me and gave me excellent direction. I utilized football to help myself become a first-generation college graduate in my family. I worked through many challenges as I embarked on this journey, and in this book, I am giving you the twenty keys that allowed me to achieve this high level of success and telling you how to apply these factors, no matter where you are in life today.

# INTRODUCTION

Growing up in a family that struggled economically helped me understand the importance of helping others build strong futures. My goal is to help others see the importance of an education at all levels. Through my educational journey, I've been able to advance to a life that's far beyond what I could have ever imagined. It takes a lot of hard work and passion to build a career that allows you to achieve success and live freely, without being burdened by debt. My goal is to help others obtain this same level of freedom. Each person can achieve this level of freedom, but it will take discipline to do so. There is a saying, "Rome was not built in one day." It is important to be patient and build a plan to achieve the things that you want to achieve in life. If you want to be an Olympic champion in the 100-meter dash, then you must create a plan to lay out how you will get there. Obviously, you cannot show-up at the Olympic Games and tell the officials that you are there to compete in the 100-meters. You must go through a process to qualify to be able to participate in the Olympic Games. This same thing is true for any goal that you want to accomplish. It all starts with the goal that you want to achieve and then you must define a plan as to how you will get to this goal. Everyone can control their own destiny, but it will not always be easy. It will take planning and persistence. You will face many challenges along this long and intense journey, but NEVER GIVE UP! Regardless of where you start in life, you can BEAT THE ODDS and be successful.

# HOW TO BEAT THE ODDS-BE DRIVEN!

**These are the twenty success factors that we will discuss within the pages that follow. Refer to this list often, and once you finish the book, create a plan to incorporate each factor into your life.**

1. Make God your foundation.
2. Develop strong family values.
3. Build a strong circle of friends.
4. Utilize education to help you grow.
5. Develop and live by a budget.
6. Overcome obstacles.
7. Learn from coaches and teachers.
8. Set clear goals and measure progress.
9. Drive to achieve.
10. Build trust.
11. Be visible to your family, friends, and co-workers.
12. Build quality character, friendships, products, and services.
13. Provide resources to help others be successful.
14. Coach and mentor others.
15. Work with others to build success (teamwork).
16. Be involved with the process.
17. Build collaborative relationships.
18. Build a stable environment so that everyone can be successful.
19. Love and care for others.
20. Repeat these items.

# A STORY *about Success*

There once was an old man who lived all alone. This man had not always lived alone. He had been married with three children. The man was now eighty-five-years old. He had lived a fruitful and wonderful life. His wife had passed away two years earlier at the age of eighty. They had been married for sixty years before she passed away. This old man and his wife had raised three children in a loving and God-filled environment. Each child was sent to college to gain an education that the old man and his wife did not have a chance to earn. The oldest child was a boy, the middle child was a girl, and the youngest child was a boy. The children were evenly spread apart in age by two years. The man and his wife worked hard and saved so that the children could enjoy a life of happiness and become successful.

One of the children decided he would not complete college, but the other two completed college. The youngest was the one that did not complete college. Instead he started working at a local restaurant. Once graduating college, the older two children started professional careers. One was a doctor and the other was a lawyer. They were very successful financially. The younger child struggled

# HOW TO BEAT THE ODDS-BE DRIVEN!

and he did not understand why he was not being as successful as the older two children. The old man sat down with him and explained the importance of an education. This was a clear example of how an education can help people to obtain a higher level of success. You can be successful without an education, but your chances for success are much greater if you have an education. Based on his discussion with his father, the younger child decided to go back to college and he obtained an engineering degree and became a very successful engineer with NASA. The old man was proud that all of his children benefited from the sacrifices that he and his wife made to allow them to get a college education. This is how you build generational wealth. Give your children something that can bless them throughout life such as God, love, guidance, and a path to an education.

*Stay focused on your goals.
Do not let obstacles or circumstances
prevent you from achieving what you
want to accomplish.*

*- Eddie Maddox, Ph.D.*

# HOW TO BEAT THE ODDS-BE DRIVEN!

# CHAPTER 1:
## *Childhood Roots*

I was born on March 6, 1964 in Anderson, South Carolina. I was the fifth child. Eventually, my parents had one more child after me, my baby sister. There was a total of six kids: three girls, three boys. I had an early start with my education, which I believe was one major factor in my success in life. I also learned to dream big as a youngster. My dreams included being a successful person who made major contributions to our society.

    As I was growing up, my parents enrolled me in a program called Head Start at a church in the small Anderson County town of Pendleton, where we lived. I was the first one in my family to attend this program. The Head Start program began in 1965, part of Lyndon Johnson's Great Society vision, but my older siblings did not have an opportunity to attend. I could tell my parents were so excited to get me enrolled in Head Start because they saw it as an opportunity to get me a head start on life and it turns out that it really worked out that way. Once I completed the one-year Head Start program, I was off to elementary school.

# HOW TO BEAT THE ODDS-BE DRIVEN!

I attended Pendleton Elementary School. I had some very good teachers who helped build upon my educational foundation. They poured much knowledge into me and I sincerely thank each of them for all that they taught me. My elementary education was second to none. The teachers really cared if I learned the materials and I was very attentive and gave my best effort to learn what they put before me. Due to my persistent attitude towards learning, I was always considered to be an outstanding student throughout my time in school, and not only that, I was also athletic. People from all walks of life gravitated towards me because I was a strong student academically and I was also a stand-out athlete, so this made me very popular with many people. I was one of few African-Americans in the school and this was consistent throughout my formative years. I battled from a childhood filled with socio-economic challenges that few people can escape and overcome to become a successful, well-educated, and high income-producing citizen. I used the challenges I encountered as a youth as a motivating force in the classroom and in sports to help me move mountains.

Based on how well accepted I was in the community, I felt like someone had favor for me. That someone was God. From day one, everything always worked out for me. I always worked hard, but I did not have to overwork myself to place myself in certain

## 1- CHILDHOOD ROOTS

positions or in order to achieve good grades or shine on the athletic field. From an early age, I had God's favor. I strongly believe that my parents' prayers and guidance helped shape my world to be such that God's favor was part of my life.

My parents prayed for all their children. One story that I was told was about a car accident I was in when I was six months old. My mother was driving the car and one of my siblings decided to grab the steering wheel as she was going around a curve that led to a small bridge that crossed over a creek. When my sibling grabbed the steering wheel, the car plunged off the road and flipped over. As the vehicle was flipping over, it ejected me out of the window and the car was only a few inches from landing on me. At the same time, I was also only inches away from being tossed in the creek that would have completely submerged me. Of course, I remember none of this, but I was told that my parents prayed that I would recover from this devastating accident. I did not sustain any major injuries from this accident, nor did it cause me any struggles later in life. Through the strength of prayer and the love of my parents, I avoided being crushed by the car and being submerged in the water as a young baby. Through their ongoing prayers, I was healed from the bumps and bruises that were caused by the accident. My parents and God were watching over me.

## HOW TO BEAT THE ODDS-BE DRIVEN!

As child number five for my parents, I always worked to make sure I was noticed in the family. I did not want to be the last one at the dinner table or the last one to get a snack. My older siblings were bigger and stronger so they had an advantage over me. Although both of my parents worked, we were a low-income family. My parents were very proud of all the kids and they always wanted to see us do better than they had done. I remember countless discussions my parents had with us stressing the importance of getting a very good education. My parents' dream was for all of their children to be well educated so that we could build families and strong futures for ourselves. My parents showed us examples of how an education could help us be successful. For example, they took us on field trips to neighborhoods where well-educated people lived. There was this one community in nearby Anderson on a golf course that I considered the type of neighborhood that I would raise my family once I was married. At the time, I was just eight years old and I already had goals of how I wanted to live. I learned during these field trip discussions that an education was my avenue to reach a successful life. My measure of success consisted of raising a family and being able to achieve the things that I wanted to achieve. My parents also taught me that it would be family more than anyone else, that will be there for us during the successful times and the low points that we may run into during our lives' journey.

# 1- CHILDHOOD ROOTS

I grew up not having anything extra, but I grew up with the desire to be a productive citizen in this country, make a positive impact on other people, and inspire them to be the very best that they can be. My parents had the same dream for me, which was why they were so excited to enroll me into the Head Start program. When I was growing up, I did not like leaving my mother's side. I was very attached to my mother, so when I went to Head Start, I would cry all day long because I wanted to be with her. The young lady who was in charge of the Head Start program would always come to rock me to sleep. She did this on a daily basis to make sure that I did not feel alone, and I think this was the beginning of the development of my character, because that young lady could have said, "You need to stop crying," and pushed me aside. Instead, she really saw something special in me that wouldn't allow her to treat me harshly. She showed a lot of compassion and love with what she did. People like this build better communities for us to live in. We all should have compassion and love for others to help them develop to their full potential.

**Our childhood often plays a very important role in who we become as adults. What were some of your early childhood influences? What do they tell you about who you are now? What odds and statistics did you beat (or are working towards overcoming) from your childhood?**

# HOW TO BEAT THE ODDS-BE DRIVEN!

# CHAPTER 2:
## *Maintain Your Focus Despite Naysayers.*

As a youngster, I was always fascinated with gaining new knowledge. I was totally consumed with trying to figure out how things worked. So, as I started school, I was totally intrigued with the fact that I was learning new things. During my time in Head Start, I built a strong foundation with numbers. I remember how we were taught how to count. I enjoyed knowing that I could use my fingers to count the number of items that were on the table. I also learned how to add small numbers. It was fascinating to me to see that when I had one pencil and the teacher gave me three more, I then had four pencils. It was enjoyable and easy for me to understand that one pencil plus three pencils equaled four pencils. The foundation was set, and I knew that I would always love working with numbers. My next step was elementary school.

In elementary school, I began to learn how to use numbers in science-related experiments. I learned that there were nine planets in the solar system (well, before Pluto was demoted) and that the sun's gravity pulls on the planets and holds the planets in alignment to revolve around it. This was exciting for me to see and understand.

# HOW TO BEAT THE ODDS-BE DRIVEN!

At the same time, on television, I watched astronauts going to the moon and returning to earth in capsules that landed in the ocean. This really kept me captivated and wanting to learn, because I too wanted to explore just as the astronauts were exploring.

By the time I completed Riverside Middle School, it was obvious that I would become an engineer because I enjoyed the classes that focused on electricity and mechanical systems. My goal was to determine how physics worked and how it allowed people to defy gravity. I was interested in automobile engines and wanted to build a better engine. The ideal career for me was engineering. I knew there was only one way I could become an engineer and that was by working hard in school and making really good grades so that I could achieve my goals by going to college.

I have only one bad memory in middle school and it was in a history class. I want to go on record and say throughout my years in school every teacher that I had considered me a good student and good citizen except my history teacher in seventh grade. All the other teachers thought I was the ideal student. This one history teacher did not show me the level of professionalism that all my other teachers showed me, and he even told me one day that I would not be successful in his class. I responded, "Sir, I will give my best effort in your class and I will make an A." Guess what? I proved this teacher wrong and I earned an A in this class. I consider this as a moment of char-

## 2- MAINTAIN YOUR FOCUS

acter building and it demonstrated a trait that is still true for me today. When someone doubts me, I do not allow it to take me off course because I understand what the goal is and my job is to meet the goal even if there are naysayers around me.

## LESSON:

Everybody will not like you, but you cannot allow that to prevent you from performing at your very best level. When you work hard and give your best, you can prove to the world that you can excel despite challenges and adversity. It is important to understand that if you work through situations and remain focused on the goal, you can overcome them even when there are individuals that may be doubting your ability based on preconceived notions that they have of you due to your socio-economic background or other reasons. Plan to stand up and make a positive mark on any situation that will be put before you even when you may have a battle on your hands. People who work hard and diligently are the ones that will overcome situations that may not look favorable. You also must possess self-confidence. I always want others to see my self-confidence as a sign that points towards success.

**HOW TO BEAT THE ODDS-BE DRIVEN!**

---

*Train your mind so that you can build a solid foundation which will support your worth.*
*- Eddie Maddox, Ph.D.*

---

# CHAPTER 3:
## *Make Education Your Foundation for Success*

When I was in the eighth grade, I had a math teacher named Mrs. Williams who looked out for me making sure I challenged myself academically regardless of the circumstances. At the end of my eighth-grade year, I had to sign up for classes to take during my ninth-grade year. My plan was to make sure that I was the best football player that ever existed on the planet, so I was going to make sure that I didn't allow my studies to hamper me. So, I devised a plan, which seemed good to me, but in reality it could have had terrible ramifications had someone not intervened. I did something I shouldn't have done. I signed up for the basic courses in all my ninth-grade subjects. I signed up for basic math, basic English, and basic level in all the other subjects.

Keep in mind, this course of action contradicted how I was performing academically. In middle school, I was at the top of the class, and additionally I was taking advanced courses in every subject. Mrs. Williams, my math teacher, obtained my schedule and she

## HOW TO BEAT THE ODDS-BE DRIVEN!

was not pleased with my "brilliant idea." As a result, she changed my course load to all advanced and college preparatory courses. What a blessing in disguise! It was another sign of God's favor. I could have messed up and destroyed my chances to go to college by taking all basic courses, but to prevent that she changed it so that I would be in advanced courses and college preparatory courses. As my math teacher, Mrs. Williams knew what I was capable of and she made sure that I did not sell myself short. I did not let myself or her down.

## LESSON:

**Never cut corners or take the easy way out. It is critical to push yourself so that you reach your full potential. Do not settle for taking basic courses in school when you are capable of being successful in higher level courses. Vince Lombardi said it best, "Perfection is not attainable, but if we chase perfection we can catch excellence."**

Throughout high school, my love of math and science continued. My high school years were equally as exciting with the higher-level math and science classes. It was a slam dunk that I would be an engineer. I was able to tear things down and build them back. I was able to mathematically figure things out with algebra, trigonometry, and calculus. I used the scientific method to carry out experiments. The level of excitement that I achieved in academics

## 3- MAKE EDUCATION YOUR FOUNDATION

was second-to-none. I thoroughly enjoyed going home every evening working through my homework. I would sit at the table and methodically work through problems and write reports and papers for my classes. I was clearly building the required workload management that was necessary for someone at the college level. I was completely convinced that I would be a college student one day, earn a degree doing something I loved, and in return would become financially self-sufficient. The driving force behind this was that I knew my parents would be proud of me. I could tell that my mother and father were so proud of my many accomplishments both academically and athletically, and I desired to make them happy. When we would go out to the grocery store as a family, other adults from the community would always come up and let my parents know how proud they were of my academic and athletic accomplishments. This kept me motivated to accomplish the goals I had set for myself.

I consider myself a poster child for how to obtain an education without paying one cent. This was a direct effect of my decision to study diligently and develop the skills that I needed to succeed in school. While I excelled on the high school football field, I obtained my undergraduate degree with a football scholarship at Western Carolina University, after spending my freshman year at Wofford College. Once I went off to college, I learned how to be independent. Through middle school and high school, I developed

# HOW TO BEAT THE ODDS-BE DRIVEN!

the study habits that were required at the college level. I continued to use them and they continued to work. I was now in the place where I could obtain the degree that would catapult me to the level of success that I was working towards. I was now able to work on the engineering courses that I wanted to take so that I could obtain my degree. I worked really hard and I had many challenging courses. I had a very good professor for statics and dynamics. This teacher was very tough as well. He gave us two questions on each exam. There were only three grades you could make on his exams. You could make 100, 50, or 0. He did not give partial credit for the problems in either class. The answer was either right or wrong. He, like all my other engineering teachers, had the students draw a rectangle around the final answer for the problem. This made it very clear what we defined as our final answers. This approach still resonates with me, because in everything I do today, I want to make sure it is clear and not ambiguous. This allows others to make clear and concise decisions and it eliminates confusion.

  Being a college football player is demanding. College athletes are required to give a lot of effort to their sport. If you are on scholarship for the football program, then you will be required to give much effort centered around football. Football in college is like having a job. Everyday we had to lift weights. Weight-lifting was required both during the season and during the off season. During the season, we were required to watch and evaluate film on a daily

## 3- MAKE EDUCATION YOUR FOUNDATION

basis and then we practiced. Each day during the season, I devoted at least five hours to football. The rest of the time, I was attending class or studying, to which I devoted seven to eight hours daily. So, my education was not really free; it required a lot of effort. I personally would like to salute all college athletes because playing a college sport requires a lot of time and effort and it leaves just a little amount of personal time. While I was in college working on my undergraduate degree, I only had enough time to go to class, study, and participate in football. There was not very much social time.

My wife, Candace, who I began dating in high school, obtained a business degree and once we were married and had children, she became a stay-at-home mom. Once both our children were in school, my wife started volunteering at the local elementary school. She was spending a great deal of her time there and seemed to be enjoying it. One day I suggested she should think about applying for a teacher's aide job. Later a teacher's aide job became available. My wife worked as an aide for many years. I saw the amount of love and dedication she devoted to her job and felt she had found her true calling. We both came to the conclusion that it's rare to find a job you're passionate about and agreed she should pursue her love for teaching. She enrolled at a local university in their teaching program and two years later, she had obtained her teacher's certification. As she was going to school, I became motivated to go back and obtain my master's degree. This was certainly

# HOW TO BEAT THE ODDS-BE DRIVEN!

a blessing from my wife to me. I started the enrollment process and the manufacturing company that I worked for paid for my master's degree and also paid for my doctorate degree. I am very blessed to have received my complete higher education without having to pay any money. Oprah Winfrey has said an education is the key, the door, and the lock. I want to add to that and say my education is the foundation, the building, the key, the door, and the lock for my success. This can work for others as well. It is an education more than anything else that can help an individual become successful.

By now, it should be very obvious to you that my education enabled me to achieve a high level of success. It would not have been possible if it were not for the foundation my parents laid for me during my early years. My father only completed school up to the sixth grade, but he was very wise. My mother completed school up to the eleventh grade. My mother was also a very bright and intelligent person. They grew up during a time when kids were required to work on the farm instead of attending school. This was required to help the family meet their financial and food needs. They both knew the importance of an education and they made sure that I had an opportunity to build a sound foundation early. My education is their education. My achievements are a symbol of what they taught me as a young man. Both of my parents encouraged me to be driven and to never give up on my dreams!

*Do not make excuses for your current position in life. I encourage you to make time to change this position in order to reach success.*

*- Eddie Maddox, Ph.D.*

# HOW TO BEAT THE ODDS-BE DRIVEN!

# CHAPTER 4:
## Identify your Leadership Qualities
### Sports help build leadership qualities

Football helped me become a leader. I started playing the game of football at age six, and I played football all the way up through college. Football was the top influence that gave me the drive toward success. As a young six-year-old football player, I did not like playing the sport because it was physically demanding, especially an exercise called "six inches." This exercise required you to lay flat on your back and raise both legs and feet off the ground approximately six inches. It made my stomach ache.

Coach Brewer was our head coach and he would tell us to hold our feet up for a timespan that seemed like days. Of course, that was not true, but this exercise really hurt my stomach, so the seconds seemed like minutes for me. Then after many rotations of holding our feet off the ground six inches in the air, the coach would tell an-

# HOW TO BEAT THE ODDS-BE DRIVEN!

other player to get up and run across our stomachs, stepping on us with their shoes. After this exercise, we would stand up and start running in place and then the coach would blow the whistle for us to hit the ground with our bellies. This made my stomach hurt even more and it was not a good feeling. I quickly developed a dislike for football.

After a few days, I told my parents that I no longer wanted to play football. I made a decision that I had a high level of confidence in, and I was pretty much finished with the sport. Due to the tough exercises, I no longer wanted to play football.

After I was away from practice for three days, the head football coach paid my parents and me a visit. He told us that he really wanted me to come back out and play football. He went on to say if there were any exercises or drills that I did not want to participate in, that I could sit them out. Little did I know the coach saw I had exceptional athletic abilities. As the quickest kid on the team, he saw I could become an outstanding running back. I had no idea what a running back was on a football team. My parents convinced me to go back to practice. So the next day after the coach visited our house, I went back to football practice. At the young age of six, I quickly realized the coach considered me a leader on the team. From there, I accepted doing those exercises, and I began to like th-

## 4- IDENTIFY YOUR LEADERSHIP QUALITIES

em, because I could see that they were making me stronger and more fit. At that point, I realized what the phrase, "No pain, no gain" really meant. If you don't have some pain through a situation, you're not going to gain a whole lot. Plan to have pain if you expect to make gains.

This incident with the coach also caused a great deal of self-efficacy within me. My own belief in my capacity to do well in football grew. I had a new level of confidence and motivation that I could be, and already was, a leader on the football field. This was a major milestone in my young life.

As a football player, I would score most of the touchdowns each game. I developed the skills of a leader and a team player. I learned how to work with other players on the team to develop plays on the offensive and defensive sides of the ball. These leadership and teamwork skills remain with me today.

One vivid memory dates back to 1972. Our Little League team was playing in a bowl game called the Dogwood Bowl. I scored five touchdowns and received the MVP trophy. The next Monday at school, my teacher recognized me in front of the class. That was a very special moment to me because I was being recognized for my football accolades at school. I had never received any football praise at school. Although I had received many MVP awards throughout

# HOW TO BEAT THE ODDS-BE DRIVEN!

my Little League career, this one was very special because I was recognized in front of my academic class. After the incidents in my first week of practice, I quickly stepped into my abilities and was on my destiny path as it relates to my football career.

## LESSON: *Be Confident*

**When you are confident, you will achieve more and perform better as a leader. Sometimes, in order to gain that inner confidence, you have to push past the pain and discomfort. "No pain, no gain."**

As I developed as a Little League football player, I realized there was a need for me to work harder to become stronger and faster. I read many books on how to become stronger and faster and I created my own special workout area in our yard. This area consisted of a big tractor tire and weights. Everyday during the off season from football, I worked out in this area to become stronger and faster. After my last year in Little League football (my eighth-grade year), I realized that I would have to play against players at the high school level, who were bigger, older, and certainly stronger. I knew I needed to turn up my level of intensity while training so that I could be prepared for the next level.

## 4- IDENTIFY YOUR LEADERSHIP QUALITIES

**LESSON:** *Seek out Yours and Others' Talents*

It is important to understand that you may experience some rough days and hard moments. You should stay focused and endure the pain because this may be the area where you are strong and it may be the area that allows you to become very successful. Returning to football practice helped me understand that I can endure pain to obtain success. My football ability would have gone unnoticed if Coach Brewer did not see the potential that I had in this game. This has made me become a person who seeks out talent, and I am sure to share my insights with the individuals when I see they have a certain ability that could one day help them be successful. The game of football gave me an opportunity.

During the summer of the start of my ninth grade year, I went out for the junior varsity (JV) football team. I practiced with the JV team and personally, I didn't feel like I was playing to the level of my ability. I always focused on working very hard at whatever I did and giving my best effort. In this situation, I was giving my best effort. I always told myself that as long as I gave my best effort, regardless of the outcome, I was satisfied. I didn't feel like I was excelling nor did I feel like I was playing to the level required to be a starter. I was very hard on myself as a ninth grader, and I always

# HOW TO BEAT THE ODDS-BE DRIVEN!

focused on continually improving. As long as I gave my best efforts, regardless of the outcome, I would be satisfied. On the other hand, if I was not giving my best effort, regardless of the outcome, I would not be satisfied. If I was not giving my best effort and I ran for a thousand yards, I wouldn't be satisfied, so my drive and method of becoming successful was to give my best effort. If I gave my best effort, I knew that I had an opportunity to be successful. My goal in life was to be the very best at what I did, so this meant there was absolutely no compromise when it came to effort. Although I did not feel I was playing to the best of my ability, I actually was playing at a great level in the coaches' eyes. Based on the coaches being excited about my performance, I was satisfied because I knew I was giving my best effort and it was being recognized.

## LESSON: *Develop a Success Mindset*

**Success requires work. If you believe in yourself, then you can move mountains and reach a high level of success. You should create goals for yourself. I highly recommend that you define your goals and write them down. It is also a good practice to review your goals in the morning and in the evening each day. A successful person with a success mindset is driven and contagious. You should work to understand what is necessary on a**

## 4- IDENTIFY YOUR LEADERSHIP QUALITIES

**daily basis to achieve the goals that you have set. This carries over into everything you do and you can become a highly successful person.**

After practicing with the JV team for approximately two weeks, the varsity head football coach wanted to talk to me. He told me that he had been watching me and he believed that I had the ability to play on the varsity football team as a freshman. I was totally overjoyed with this discussion, but I told him that I needed to review this with my parents to make sure they were okay with me moving to varsity. I spoke with my parents and my father was A-OK with me playing on the varsity team as a freshman, but my mother was apprehensive because she felt I was too young and I'd get hurt playing with twelfth graders as a ninth grader. We thought about this decision for several days and we weighed the positives and negatives of me playing on the varsity team as a freshman. It turned out that there were more positives than negatives, so I notified the coach that I'd be playing on the varsity team.

# HOW TO BEAT THE ODDS-BE DRIVEN!

## LESSON: *Create a Decision-making Process*

As you journey through life, you will come upon many decisions that you will need to make. It is wise for you to create a process to use to help make the best decisions. I always work to define a list of positives and negatives associated with a decision. If the positives outweigh the negatives, then typically I go with the decision to go forward. In the example above, I played varsity football as a freshman and my decision to play was based on weighing the positives and negatives. My son Brian was very fortunate as a high school football player. He had scholarship offers from most of the major college football teams. We used this same process for him to narrow his top five schools down to one. This process works, and it proved effective for both my son and me.

We started the first day of high school varsity practice by stretching and doing calisthenics. Calisthenics are exercises that we performed prior to practice. These exercises work large muscle groups and include jumping jacks, push-ups, and sit-ups to name a few. The next thing we did was a drill called "county fair." I had no idea what county fair was, but I could tell that the majority of the returning varsity players did not want to do this exercise. During the county fair period, the team split up into five groups. It consisted of five

## 4- IDENTIFY YOUR LEADERSHIP QUALITIES

stations, and at each station you did exercises that pushed your bodies to the limit. Once we finished the exercises, we'd go on to the next station. The exercises at each station were fundamental skills that really put the body to the test. These drills worked on the conditioning of all the players.

After the first station I was physically exhausted, however, I was able to get through all the stations with no problem. I was physically ready to play at the varsity level after I saw that I made it through the county fair with no issues. Once we finished, we worked on offensive and defensive drills.

I was a running back on the offensive side of the ball and I was a defensive back on the defensive side of the ball. I enjoyed playing offense the most because I could score touchdowns. As practice progressed, the coaches started to put me in the rotation with the starting offensive and the defensive teams. By the time the season started, I was alternating as a running back on the first team as a ninth grader. I was really proud of how I had progressed. This accomplishment came from hard work and teamwork. By season's end, I was also starting as a defensive back on the defensive side of the ball.

## HOW TO BEAT THE ODDS-BE DRIVEN!

The defensive back position was one that gave me some challenges. In the first game that I started in this position, I had to defend a couple of fast and tall receivers. I was five feet seven inches tall and the receivers that I was playing against were six feet three inches tall. These receivers caught passes over me throughout the whole game, and I physically could not stop them, even though I was definitely giving my best effort. I thought that this would cause me to lose my defensive back position, but our head football coach told me that both of those receivers were major college prospects and that they would probably play in the NFL, so it was not an issue that they caught passes and I was not able to defend them. The coach was proud of the effort that I gave even though they were catching the passes because I was still able to tackle them and not allow them to score. Coach told the team that I was an example of the effort that he wanted to see on every play by every player. I was in the ninth grade and I was a starter. It was almost like the situation was unreal, but I have always had God's favor and I've always been a hard worker. My approach to whatever I am part of is to always give my best effort. It is very clear that if you work hard, things will work out for you.

## 4- IDENTIFY YOUR LEADERSHIP QUALITIES

In the off-season, I continued to work extremely hard in my yard to get stronger and faster. Throughout my high school career, I was all-conference every year, an all-area and an all-state player. Fortunately, I graduated sixth in my high school class out of 210 students, and I received a football scholarship to Wofford College in Spartanburg, South Carolina. Football, determination, and drive placed me in the spot to become the first college graduate in my family.

As a freshman at Wofford College in 1982, the coaches had a plan for me. I came in and played right away as a freshman. I also became known as Little Hershel Walker. Hershel Walker was a great football player at the University of Georgia. He was an extremely talented running back, probably the best college running back ever. He wore number 34, and I ended up getting the same jersey number. I didn't pick that number; it was what they gave me, and of course the "Little Hershel Walker" thing took off because that was his number and I was extremely fast like him. I had a very successful freshman year at Wofford as a player and as a student. However, I decided I wanted to get a technical education, but Wofford College was a liberal arts school. It was the only school that had offered me a scholarship when I came out of high school, so that is where I landed.

# HOW TO BEAT THE ODDS-BE DRIVEN!

I had a friend who was at Western Carolina University (WCU). My friend told me to contact the coaches at the university. I made contact with some of the coaches there, and they said, "Hey, we'd like to get you up here." I transferred to WCU after my freshman year. In 1983 I had to sit out of football one year because I had transferred. However, I still had to practice with the team during that season, which was good. It kept me in shape, and I was able to keep my football moves and play against the competition. In the spring of 1984, we had spring football practices, so I was now ready to show what I could do. During spring football practices, players earned and secured our positions. I was eligible and participated in the spring football practices. I had an exceptional spring football session, and I proved that I was the best running back on the team at that point. I walked away from spring practice as the starting running back for the WCU Catamounts. I felt like I was standing on top of the world. Here I was as a transfer student who jumped right in there to get my starting position.

Once I returned to Western Carolina in the summer of 1984, I found out from the coaches that they had recruited and signed four new running backs. When I came to summer camp, I was now listed as the fourth running back on the depth charts. So, I went from a starter to number four and had not played one minute of football

## 4- IDENTIFY YOUR LEADERSHIP QUALITIES

for the school. Simply by stats, they thought these running backs just coming out of high school were better than me and they had never played a down of college football or spring practice nor had they practiced with the team. The coaches had put three of the running backs who were signed in front of me. I was very calm throughout this situation and asked myself how this was possible when they had not played one down of Catamount football. I was determined that I was not going to give up. I'd give my very best. That's my motto. I was going to give my best regardless of the outcome.

The first game of the season came around. We played at Boston College, and Doug Flutie was the starting quarterback at Boston College. He was a very good starting quarterback at the time. I was excited to play against him, but the sad part is that I did not get to play one down of the game. I couldn't understand why I wasn't given an opportunity to play. Two more games went by and I still did not get to play. I did not touch the field for the first three games. At this point, I decided that I needed to do something.

On the Sunday after the third game, I decided to speak with the coaches. I was very respectful when I told the coaches, "I know that I am the best running back on the team." For some reason, they were playing these other guys and were not giving me a chance. The

## HOW TO BEAT THE ODDS-BE DRIVEN!

coaches told me that they didn't think I was as good as the other players they were playing. Of course, I didn't agree, but I reminded myself why I was in college. I was there for a college engineering degree, and regardless of what happened, I would give my best in football and in my classes, and at the end of the day, I was getting my education paid for from football. That was how I was wired, and it gave me peace to know that I was giving my best effort. As long as I was giving my best effort, regardless of how the circumstances work out, I'm at peace with the situation. There are some things as individuals that we cannot control. I knew I could not force the coaches to let me play. The only thing I could control was my effort.

### LESSON: *Prepare and Persevere*

**It is important to believe in yourself. If you do not believe in yourself no one else ever will. There will be moments when you may feel like the world and everyone around you have let you down. This is not the time to have a pity party, but it is the time for you to stand up and take charge. I highly recommend that you talk to others to understand why they've made certain decisions, when the situations are not going how you would like for them to go. Regardless of what you are told, you must make some key decisions about how you will proceed to overcome the**

challenge that is before you. Be prepared at all times, because you do not know when you may get your chance. You stay prepared by working hard to make sure that you are on top of your game.

Week four of my first season at Western Carolina came around. I practiced all week and didn't get any repetitions with the starting offense during the week of practice, which is to say that I wasn't going to play during the game. We were playing against Virginia Military Institute (VMI). Once we arrived at VMI, the running back coach came up to me and said, "We decided that we are going to let you start the game today." I was shocked because they hadn't let me play up until this point or even get repetitions at practice with the first team. I didn't understand why they were going to let me start when I didn't have any plays with the first team during practice. It really didn't make sense to me, but I'm the type of person who will give my all regardless of the situation. I was going to take advantage of the opportunity, because I didn't know when I would get another chance. I would give it my best shot. My approach is to make sure I'm always ready, so even during the game, although I hadn't practiced or prepared with the team, I was mentally and physically ready to be tested.

# HOW TO BEAT THE ODDS-BE DRIVEN!

The game started and VMI kicked off the ball. Our offense was on the field first. The very first play was a running play off the right tackle, and I had the ball in my hand. I went for twenty-five yards on the very first play. On the second play, I picked up twenty yards on the left side. The very next play, I was handed the ball from the quarterback and I scored a touchdown from the fifteen-yard line. I ended the game with four touchdowns and a 195-yards rushing. It was very clear that I now was the starting running back on the Western Carolina University Catamount team.

It also became very clear that whether we are working somewhere or playing a sport, we must stand up for ourselves when we believe in something. It doesn't matter if it's a sports team, your job, or something in society that you're a part of. If you believe in it, stand up for yourself otherwise people will pass you by. If I had not talked with the coaches I probably would not have played one down at Western Carolina. On top of that, I had a season record for consecutive 100-yard plus games. I received Academic All-American honors for the 1984 and 1985 seasons as a football player.

## 4- IDENTIFY YOUR LEADERSHIP QUALITIES

When I graduated in May of 1986, I had another year of eligibility to play football. I decided that I had a better chance of being a productive citizen in the working world than being a successful player in the NFL. I had some chances to try out on a few NFL teams, but I decided to say goodbye to football as a player and go into the working world. I started my professional career in manufacturing that month.

**HOW TO BEAT THE ODDS-BE DRIVEN!**

---

*Do not feel sorry for yourself when things are not going the way you want them to go. You should decide to use these situations as motivation to grow.*
*- Eddie Maddox, Ph.D.*

---

# CHAPTER 5:
## *How to Overcome Obstacles*

**Life will throw many challenges at you, and you will need to stay focused to make it through them successfully.**

During my senior year of high school, I was having an outstanding year as a standout football player. I was leading the state in rushing touchdowns and rushing yards. I was averaging fourteen yards per carry, and honestly, I could not have written a better script for what I was experiencing as a senior football player. There was a major turning point in my life and athletic career that happened during the ninth game of the season, and I still remember it like it was yesterday.

    We were playing the Liberty High School Red Devils. I was having another stellar night with over 150 yards rushing and multiple rushing touchdowns. I had just scored a touchdown. After the touchdown, we lined up to kick the extra point. This play seemed like it happened in slow motion. As a running back, I lined up on the left

## HOW TO BEAT THE ODDS-BE DRIVEN!

side of the formation with the side of my right foot next to the back of the offensive end's foot. The ball was snapped, and the offensive end fell on my right knee. I felt excruciating pain and I could not get up to walk off the field. The coaches and trainers came out to see me and worked on my knee, but it was obvious that I was hurt really bad. Several players had to help me off the field. This was a painful situation because I saw my football career flash in front of my eyes due to this knee injury. I now had a major challenge to deal with. Eventually, I had surgery to repair my knee. It was now up to me to deal with the challenge of getting my knee back to full strength. I worked really hard from January to July so that I would be ready for my introduction to college football. This was a situation I wished never occurred; however, it was a test of my will and devotion to overcoming an obstacle.

After having surgery to repair my knee, I developed a major physical therapy program with the local orthopedic surgeon who had performed the surgery on my knee. The plan was filled with stretching, running, weight-lifting, and therapy. I worked out on a daily basis. I ran 100-meter sprints. I ran up and down hills (forty-yard sprints). I lifted weights in the high school weight room, including leg weights and squats, and I went through controlled physical therapy through the doctor's office. They had some state-of-the-art equipment for resistance training. The harder I pulled down, the more resistance it would give, so it really maximized the level I was pushing to get me back stronger. I did physical therapy the entire summer, so I was able to come back full speed when practice started for my freshman football year at Wofford College. I was also able to get in the running back rotation at Wofford. I was able to cut and run with no sign that I recently had a knee injury.

**LESSON:** *Never count yourself out.*
*There is always a way.*

    **I successfully dealt with this injury and came out of it stronger and faster. This injury gave me the desire to fight when things were at a low point. Each of us will reach low points, and it is up to us to fight and overcome them. It starts with a strong belief in God and having the courage and heart to work through a situation even when the odds may be against you. Never give up! You can overcome challenges that may be thrown in front of you. Once you get through the challenge, you will look back and tell yourself that you can overcome obstacles that are thrown in front of you. It comes down to hard work and the desire to never give up.**

# HOW TO BEAT THE ODDS-BE DRIVEN!

We are all sometimes put in places that we do not want to be in and these are challenges that we should face with an attitude of determination. As you have read, my goal in life is to make sure I give my best effort and things will work out positively. You must be happy with where you end up in life. You are special, and it is up to you to make sure that you understand you are special. Living your life to the fullest starts with never giving up. Oprah Winfrey said, "Challenges are gifts that force us to search for a new center of gravity. Don't fight them. Just find a new way to stand." I have faced many challenges over the years. I refused to give up on my dreams. I encourage everyone to dream big and determine what work is required to achieve the dream. When I face a challenging situation, I analyze it and determine two things: the benefits I can get from the situation and the lesson it is teaching me. Determine the benefits and lessons of every situation you encounter.

Another major challenge I faced occurred on a Friday afternoon when I was in my early forties. At this time, our children were fifteen and ten-years-old. I had a co-worker who wanted to pick-up some masonry blocks I had on a piece of property I had just purchased. My co-worker and I went out and started placing the blocks in his truck. Any time that I do work such as this, I wear safety glasses. I had my safety glasses on while we were moving the blocks.

# 5- HOW TO OVERCOME OBSTACLES

I bent over to pick up a block and turned around into a tree branch that hit me on the right side of my face near my eye. Thank God I had on my safety glasses! We finished putting the blocks on the truck, and I went inside the house, took a shower, and had a bite to eat. Shortly after eating, I watched television for roughly an hour and then decided to go to bed. I woke up on Saturday morning and right away I noticed that my right eye was blurry. I stayed in bed for a while because I figured that I was just sleepy. I woke up a little later and my right eye was still blurry.

At this time, I was in denial that something was wrong with my eye, so I stayed in bed pretty much all day and went to sleep early on Saturday night. I woke up on Sunday morning and my vision was the same as it had been the day before. I told my wife I needed to go to the urgent care facility. The doctor who saw me said I needed to get to an ophthalmologist to look at my eye right away. The ophthalmologist that I was referred to came into his private practice on a Sunday and looked at my eye. He gave me some very bad news. He said, "Mr. Maddox, I hate to tell you this, but you have some permanent damage to your central vision." The blood supply had been cut off from my eye for a short period of time and this had caused me to lose the central vision in my right eye. I was devastated; however, I had to work to determine why the blood supply was cut off for a short period of time.

# HOW TO BEAT THE ODDS-BE DRIVEN!

The next day, which was Monday, I visited a specialist who performed a medical test to see if I had any blockages. I did not have any blockages that might have caused this situation. I also visited my regular physician. I spoke with him and started trying to connect some dots. I mentioned to him that I constantly had sinus pain on the right side of my face and now I had lost my central vision on the same side. I told him that he should run some tests, including a Magnetic Resonance Imaging (MRI), to determine if there was something going on that may have caused both situations. The doctor was convinced that there was not anything abnormal happening. I was very persistent, and he finally said he would order me a Computerized Axial Tomography, better known as a CAT scan. A couple days later, the doctor called me and said the radiologists reading the scan saw something that they had to take a closer look at to be sure there were no issues. He scheduled an MRI for me. I completed the MRI right away and the doctor called me later that day and told me that the MRI showed that I had a pituitary adenoma. This is a nice way to say that I had a pituitary tumor. This is a growth that grows in the pituitary gland in the skull, which is below the brain and above the nasal passages.

## 5- HOW TO OVERCOME OBSTACLES

When the doctor called me, I was driving home from work. This news floored me. Once I hung up the phone, I started crying uncontrollably. I cried for twenty minutes which was the amount of time I had to get home. I had no idea what this news meant, but I was worried about my wife, daughter, and son. "Do I have cancer?" "Would I survive?" "Could the tumor be removed?" These are all questions that roamed around in my head. My job is to take care of them, and I had no idea what this news would lead to. Once I got home, I composed myself and said that I would meet this challenge head-on just like I had met other challenges.

## LESSON:

**Give yourself a moment to process bad news, but don't allow the bad news to paralyze you to doom.**

Over the next few months, I had to visit many doctors to understand the situation. I visited my general physician, neurologist, and endocrinologist. All three of them communicated with each other and me to make sure that we fully understood and treated this situation properly. After much consultation, I decided that the neurosurgeon would perform surgery to remove the pituitary adenoma. The endocrinologist wanted to perform a few more tests before a final decision was made about surgery. He received the results back and

## HOW TO BEAT THE ODDS-BE DRIVEN!

determined that the pituitary adenoma was a prolactin-producing tumor. This type of tumor is benign, and it can be treated with medication. The neurologist saw these results and told me that my best option would not be to operate, but instead to treat it with medication. I agreed. This challenge was met head-on with expert knowledge.

## LESSON:

**Make sure that you have factual information for anything that you face so you can make a decision that will take you in the right direction. Challenges help make us stronger and they allow us to learn and grow.**

The next challenge that was put in front of me was a major situation. As a football player, I worked out and became well-conditioned while I played football. Being in good physical condition has always been a top priority for me. My physical condition was so good that my resting heartbeat rate was thirty-five-to-forty beats per minute. The average person's heartbeat is sixty beats per minute. My heart also skipped a beat every eight beats. This was normal for me and had been the way my heartbeat had always been. My general physician told me that we would need to keep an eye on this since my resting heartbeat rate was thirty-five-to-forty beats per minute.

## 5- HOW TO OVERCOME OBSTACLES

In 2018, when I was fifty-four-years-old, I was on my way to pick up my wife from the airport. I left work and got about five miles away from the facility and everything started spinning. I was so dizzy that I had to pull over on the side of the road. I was determined to go pick up my wife two and a half hours away so I let the spinning subside and kept going. The spinning started again shortly after I took off in the car. I quickly realized that I was not going to be able to drive to pick her up. I was finally able to arrange to have my wife picked up and she and I made it to the doctor. He told me that I immediately needed to go to the hospital and get a pacemaker inserted to raise my resting heartbeat rate to sixty beats per minute. I received my pacemaker and it functioned fine. Through this situation, I utilized the knowledge of a professional to make the right decisions.

## LESSON:

**Surround yourself with people who are knowledgeable and can help you make decisions that impact your life. Relationships matter.**

# HOW TO BEAT THE ODDS-BE DRIVEN!

# CHAPTER 6:
## *Achieving Financial Success*

Over the years, I have learned that successful people build strong relationships. You should work very hard to build good relationships with family, doctors, preachers, friends, and others. Also, work to coach and mentor people in these relationships. There is enough money for everyone reading this book to become a millionaire. I believe that we should live so that others will want to be like us and as role models we give others something to strive to be one day. I intentionally share all the things that God has allowed me to achieve at the level I am at with my family, friends, and during discussions and/or speeches. I want every person to live a life that is free from debt. It is critical to be passionate about the things that you believe in because they will help you get to a debt-free life. It starts with having self-confidence and believing in yourself.

Success can be defined many different ways based on the different areas of your life. Most people equate success to the financial status of a person and the things they have acquired through th-

# HOW TO BEAT THE ODDS-BE DRIVEN!

eir financial means. Achieving financial success is certainly an area in your life that you should manage so you can adequately provide for your family and help others. As a young boy, I enjoyed going to school and playing football, but I also had a goal of being financially independent one day. This caused me to really analyze the things that would be required for me to achieve financial success. Of course every person wants to be financially secure, the problem is many people do not want to follow a process and be patient to achieve financial success. At an early age, I learned the power of saving and investing to build more wealth.

After graduating from college, a young man told me to make sure that I invested in the 401k program at my job. This person told me that if I invested early and regularly that I could have over $1 million in this program after twenty to twenty-five years. Being a young college graduate and a person that understood the importance of investing, I jumped into the 401k program and I started out investing six percent of my income. Each year I increased the amount I invested until reaching the maximum allowed that I could contribute each year. This was one of the best decisions that I could have ever made because this has created a sound retirement fund for my wife and me. My wife has elected to invest in her 401k in the same manner. I will next discuss with you critical areas that you can focus on to create financial success, but let me pause here to talk about how important a successful marriage with both partners invol-ved in the budgeting is to financial success.

I saw a statistic that said seventy-five percent of all millionaires have been married at least thirty-two years. My wife and I have been married for over thirty-five years, and financially, we are within the top one percent of earners in America.

Today, after all of those years, my wife and I have a goal to save and invest seventy-five percent of our income each month. However, we were nowhere close to this target when we first got married. It took years of discipline, hard work, and focus.

She and I married during my senior year of college, and this is when we learned the budgeting process. I worked and saved a large percentage of my income from a summer engineering internship program. This money was used, along with the income my wife made during a work study program at Western Carolina, for us to live on during our first year of marriage. We defined a budget and we lived by this budget to help us get through my senior year at Western Carolina. We were successful with the budget and have used a budget to live by since that time.

# HOW TO BEAT THE ODDS-BE DRIVEN!

During my senior year at Western Carolina, we did not have very much, but we were happy because we were on our own and we controlled our destiny. A treat for us during this time was going out to Hardees and ordering a ninety-nine-cent hamburger. Having to live on a budget was the best thing that could have happened for us financially. This financial foundation has allowed us to become debt free, and it can be used by others to do the same. I have found that most people that are financially secure refuse to take on debt.

## LESSON:

**You must be satisfied with what you have and can afford at the time without borrowing money to get it.**

It is critical for individuals to have a solid understanding of their finances. Financial understanding needs to happen as a child develops. It is a very prudent approach to start your children out around age six on financial management. A young kid should understand one fact of financial understanding, which is to not spend everything that you have. The basics of financial understanding consist of five steps:

1. Create a budget.

2. Do not spend more than you earn and create multiple streams of income.

3. Save a percentage of your income.

4. Invest a percentage of your income and eliminate debt.

5. Invest in others.

These steps are pretty universal and you can find them defined many different ways by many people.

**The first step is to create a budget.** Creating a budget requires you to understand all the expenses that you have to pay each month. This budget must include 100 percent of all your expenses and what you plan to save. I have a sample budget at the end of the chapter that you can adapt to your own needs. There is an expense I have included in the budget that I personally do not recommend, and it is a credit card. I have listed it because many people do have credit cards. Eliminating credit cards from your budget should be a goal. An idea of the type of expenses that you should include are at the end of this chapter, but other expenses may need to be included based on your personal situation.

It is necessary for you to understand all of your expenses, and you must include all these items in the budget. A monthly budget is typically ideal, but you may want to have a weekly budget. It is totally up to your preference and what works best for you. The key

## HOW TO BEAT THE ODDS-BE DRIVEN!

is you must have a budget. You also must be disciplined with the money you make each month. This leads us to the second step of this process.

The second step is you must not spend more than you earn. If you earn $3,000 of take-home pay each month, then your expenses for the month must be below $3,000 per month. In the example of monthly expenses at the end of the chapter, you must make sure that you account for all your expenses each month even if the expenses are not due until the end of the year. Property taxes are typically due once per year. It is prudent to break this expense up into twelve months and save each month for this large expense. Also under this step, you should determine how to create multiple streams of income. How many companies have only one product type that they sell? The point that I am making is companies have multiple streams of income through the diversification of the product types or services that they provide to a customer. As an individual, determine something that you are good at and enjoy in order to create another stream of income in addition to your regular job. An additional stream of income could be as simple as selling shoe strings online to customers. You will be amazed at the opportunities that exist to create multiple streams of income. Just be sure you do not overburden yourself while creating multiple streams of income.

# 6- ACHIEVING FINANCIAL SUCCESS

The next step in this process is savings. You should work with a financial advisor to determine an ideal amount for you to save each month. A rule of thumb is to save at least twenty percent of your income. As noted earlier, my wife and I save 75% of our income each month. This means if you bring home $3,000 each month, then you should save $600 each month, which is 20% of $3,000. You should build an emergency fund in your savings so that you have six to twelve months' worth of money saved to cover your expenses. This emergency fund should be used only in the event of an emergency such as a job loss due to a business closure or crises. After you have built a solid savings plan, you are ready to move to the next step in the process.

Next, you want to invest a percentage of your income to help generate additional money. A financial advisor can be used to help make these decisions or in many cases individuals decide where to invest their own money. This decision is in your hands as to how you want to invest in order to make your money work for you. Most companies have 401k plans that allow you to have an opportunity to invest as well. At the same time, you should create a plan to pay down your debts so that you can eliminate them. I learned at an early age to make sure I pay for things in full and not finance them. You will incur a large amount of interest by financing items. Your goal should be to stay away from paying interest. This approach can help you save more and eliminate your debt. If you need an appliance such as a washing machine, dryer, or refrigerator, then pay cash for these items and do not finance them. You will be charged in most cases a tremendous amount of interest.

The last step of this process is to invest in others. When you invest in others, you are planting seeds that require you to continually "water" them (through future discussions) so that they grow and bear the fruit of success. There are many ways to achieve this goal. Many people give to charitable organizations such as churches and non-profit organizations. Many people also volunteer their time to help others in the community through programs like Habitat for Humanity, United Way, and Red Cross. There are many places where you can give money and/or time to invest in others. It is also critical that we invest in children. My goal is to spend time around high school athletes and motivate them to give their best efforts in the classroom and in their athletic sport(s). The children represent the future leaders of this country and world.

# 6- ACHIEVING FINANCIAL SUCCESS

This five-step process to help you move towards financial independence works very well for an individual who is not already spending more than they earn. If you are in a situation in which you spend more than you earn, it will be necessary for you to take a hybrid approach. You will definitely need to pay down debts or eliminate expenses before you can follow the savings process recommended. In some cases, you may need to create another stream of income by getting a part-time job. Even if you are able to comply with the process, a second stream of income, as noted earlier, can help you eliminate debt and become debt free.

*Remember, the ultimate goal is to become free of debt.*

Below are sample budget items. Create a spreadsheet with a list of your expenses, when they are due each month, and the amount of each expense.

| **Expenses** | **Expense Due** | **Amount of Expense** |
|---|---|---|
| Savings | | |
| Food | | |
| Water | | |

# HOW TO BEAT THE ODDS-BE DRIVEN!

| Expenses | Expense Due | Amount of Expense |
|---|---|---|
| Electricity/Gas | | |
| Mortgage/Rent | | |
| Homeowner Insurance | | |
| Taxes on the Home | | |
| Car Payment | | |
| Gas | | |
| Car Insurance | | |
| Car Taxes | | |
| Mobile Phone/Phone | | |
| Internet | | |
| Cable/Satellite TV | | |
| Credit cards | | |
| Monthly Entertainment – restaurants, movies, etc. | | |
| Clothing | | |
| School Expenses | | |

*You are responsible for your own life. Your decisions will impact your level of success. It is up to you to define where you want your journey to go.*

*- Eddie Maddox, Ph.D.*

# HOW TO BEAT THE ODDS-BE DRIVEN!

# CHAPTER 7:
## *Honor Important Relationships*

### *Family*

My mother and father taught all six of their kids that family is important. We always ate breakfast and supper together every day. This was the time we all relaxed and talked about the start of the day and how the day went and discussed any lessons that the family needed to hear from our parents. We were very disciplined with the time that we had our meals and this was great family time. This discipline with time is part of my foundation today. My parents would stress the importance of an education and they always made sure that we were not mistreated at school by teachers or other students. We were in public school during the early years of integration. This was an intense time and a defining moment for our country to allow all Americans equal access to education. Family is the critical piece to the success of people. Your family will be there when you hit the low spots in life. Your family will work to help you come out of these low spots. As I was growing up, I always wondered how I would me-

# HOW TO BEAT THE ODDS-BE DRIVEN!

et the right woman who had a caring heart like my mother had. My mother was the nicest and most caring person in the world. She gave and cared for our family unconditionally.

I was in the eighth grade and this girl named Candace Walker was in the sixth grade. When I first met her, I wanted her to be my wife. At that time, I was too shy to even say much more than "hello" to her. It was not until I was a senior in high school that Candace's uncle who was in my health/PE class asked me if I would sponsor Candace in the homecoming game. I saw this as an opportunity to finally get a date with this young lady who I wanted to be my wife. So, I sponsored Candace in the homecoming court and we started dating. The rest is history.

After being married for five years, my wife and I had our first child, a boy that we named Brian Justin Maddox. Prior to Brian being born, I would read books to him as I touched my wife's stomach. Once he was born, my wife and I read books to him daily. Brian was a very well-mannered kid. He learned how to respect others at an early age.

After Brian was five years old, my wife and I had our second child, a little girl named Kristin Nicole Maddox. Kristin was also a well-mannered child and she also learned how to respect others at an early age. My wife and I were determined to raise our two children

## 7- HONOR IMPORTANT RELATIONSHIPS

so they respected other people, especially adults. Brian and Kristin were both good students in school and they both had special talents. Brian had outstanding athletic ability and Kristin was outstanding in theatre. Brian was an outstanding football player and was highly recruited by most major colleges. He had football scholarship offers from Florida State, Clemson, the University of South Carolina, University of Florida, Louisiana State University, Notre Dame, and many others. He eventually signed and played at the University of South Carolina. Kristin continued theatre throughout much of her high school days and she continued to perform very well in plays. She attended Clemson University. I was the first generation in my family to graduate college and both of our kids continued the process and became the next generation of college graduates. Brian graduated from the University of South Carolina and Kristin graduated from Clemson University. They both utilized the process of giving their best efforts all the time and they both graduated in three and a half years.

As the kids were growing up, we did everything as a family. We ate all meals as a family and had many sound family discussions at the dinner table. Both of our kids knew the best way to be successful was to get a college degree and start a career. They admired the professions that my wife and I had and were eager to go to college and create professional careers for themselves.

# HOW TO BEAT THE ODDS-BE DRIVEN!

Brian married a wonderful woman named Alisa in June of 2018, and we welcomed a new addition to the Maddox family on February 28, 2020, our first grandchild. She was born at 7:40 p.m. Our beautiful granddaughter weighed five pounds and fourteen ounces. This was an outstanding moment for my wife and me. My granddaughter's name is Everly Annalise Maddox. This is the start of the next generation of our family, and I pray that God blesses her as she has blessed us.

Family is critical to being successful. My goal has always been to make my family the most important priority to me, and Candace feels this way too. This meant that I would not allow my job or any other personal aspirations to get in the way of time I could spend with my wife and children. They are the ones who deserve my full attention and they have always received it. I refused to allow my job to take all of my time while I failed to spend quality time with my family. We were responsible for making sure we raised our children in an environment that would give them an opportunity to be successful. This environment started with love and attention for them. We made sure they gave their very best in school from the time they started kindergarten. They both gave their best effort in school and they both were involved with extracurricular activities where they gave their all as well. We also taught our children to be respectful to others. We taught them to be honest, trustworthy, team players, to be successful and independent, to have good character, and treat people the way they want to be treated. We also taught them to be nice to others regardless of the circumstances. We taught them that the world did not owe them anything and it was up to them to create opportunities. Both of our children are independent and have outstanding professional careers.

## LESSON:

**As parents, we should teach our children the right morals and values and pray that they make the right decisions.**

## *Friends*

It is necessary to have friends in this journey of life. We cannot do life alone, and we all require friends to help us get through the long winding roads that we must navigate in order to be successful. I make sure I communicate and collaborate with my friends on a regular basis. I have found that friends can help keep us on the right road as we take on this long journey toward success. Friendship is mutual. You will find you can also keep your friends on the right road. It is a blessing to have people that you can call or can stop by to see in order to share thoughts to make sure you are headed in the right direction. True friends will tell you the truth and not tell you what

# HOW TO BEAT THE ODDS-BE DRIVEN!

you want to hear. I have friends I know will tell me the truth about a situation. I do not need someone to tell me what I want to hear because it may be convenient for me, but I want friends who are willing to have those hard conversations and will tell me when I am wrong. There are six key areas that will define a true friend. They are as follows:

1. A friend is honest.
2. A friend is trustworthy.
3. A friend will give you the shirt off their backs.
4. A friend is happy for your success.
5. A friend is a confidant.
6. A friend is reliable.

Honesty is so important in any relationship. An honest person is free from fraud and deception. I have found that it is critical to deal with people that will be honest. My friends will tell me when I am right and they will tell me when I am wrong. If someone agrees with everything I say, then I become suspicious if they are really my friend or not. In order to be successful in this world, we must improve on areas where we are weak. An honest person will point those shortcomings out to you and not allow you to think you are better than you are. Honesty shows that you are authentic and it is a reflection of your true feelings. I equate courage to honesty. It takes a courageous person to stand-up and be honest about a situation. It

## 7- HONOR IMPORTANT RELATIONSHIPS

also shows that you care, and when others see that you care, they will help you on this journey towards success. Honest people attract other honest people in their lives. This is critical because successful people can reach great heights in life if honest people surround them. Honesty leads to trustworthiness. Honesty and trustworthiness are two characteristics that are required for a true friendship.

Trustworthiness happens when you can believe in your friend and can rely on someone to follow through on a promise they made to you or others. A trustworthy person is consistent in what they do and how they handle situations. Trustworthy people have a high level of integrity. They will do the right thing, even when no one is looking. They are also kind and compassionate. They are able to put themselves in the other person's shoes and totally understand what that person is feeling. Trustworthy people are present and available for others. They will go out of their way to make time for you. If your friend tells you that they will be somewhere at 5 p.m, you can trust that this friend will be there at 5 p.m. I have many friends I can trust and that means the world to me. It allows me to be transparent which then allows me to learn and build sound strategies that can be executed. This includes business strategies and friendships.

# HOW TO BEAT THE ODDS-BE DRIVEN!

A friend will give you the shirt off his back if you need a shirt. They would expect the same of you. Friendships are very special and must be nurtured, and this can help your chances for successful outcomes. I speak with my friends at least once per week and sometimes once per day. It is good to discuss things with others to make sure your perspective of a situation is not being interpreted wrong. My wife and two children are both my family and my friends. I can count on them to be there for me regardless of the situation. My children will give me the support needed to overcome any situation. They both are grown now, but they have the utmost respect for me, and they show me through their words and actions. I watch as they make decisions and go through that process. I see them from a distance taking the scientific approach with their decisions. If I need them to come help me with something, then they are willing to drive three-to-four hours to make it happen. They are truly what you want to see in family members and friends. Having friends impacts your psychological approach and allows you to work with others more effectively. I have other friends that I associate with and in all cases, they help me understand the true meaning of a friend. My friends will aid me with information, by listening or whatever is needed at a particular time. I consider these to be true friends. I am

## 7- HONOR IMPORTANT RELATIONSHIPS

certain that if I was not financially successful, and I needed money to pay an electric bill, or another bill, each of my friends would gladly step up to help me out. It is a great feeling to know you have someone in your corner who is willing to help you.

A person who is happy to see you achieve success is a person who is in your corner. Friends show their happiness for you when you enjoy successful moments. My circle of friends is truly happy to see me succeed both professionally and financially. I am also willing to share the secrets to my success so that others can do what I have done. We live in a great country and this country offers many amazing opportunities. But in seeking these opportunities, there are many challenges that you will have to overcome. Through these challenges, it is necessary to have others in your corner in order for you to achieve a certain level of success. My success story has been made possible because of the friends and family that I have in my corner. My friends know exactly who I am speaking of because we talk on an ongoing basis and my success is their success. There are enough opportunities for everyone and my goal is to share the knowledge and experiences I have gained so others can obtain a high level of success. There is no reason that anyone should end up broke, busted, and disgusted in America today.

# HOW TO BEAT THE ODDS-BE DRIVEN!

It is critical to be able to talk to someone about sensitive information and not have to worry if others will know your business, which is why a true friend is also a trustworthy confidant. We all run into situations where we must talk to someone about a difficult situation, and having a friend to be a confidential listener is a blessing. Friends can help you win. I am fortunate to have great friends that are willing to hold me accountable and I am willing to hold them accountable in our daily walks with God.

A friend will be reliable. Friends are people that you can count on to follow through on promises that they make to you and others, and they expect the same of you. Be faithful to others and only say things you will honor. Be a great friend to others and this can help the person grow and develop. Friends guide and aid others in a manner that allows a person to become more productive and build a legacy of being a servant leader.

My wife and I are the best of friends. We make sure that we take care of each other in all categories emotionally, physically, spiritually, and financially. Every dollar that I make is considered my wife's money, and every dollar that she makes is my money as well. We have joint bank accounts, and we are transparent when it comes to money and our financial health. We have gone from zero to hero financially all because we work together to build a solid foundation. We know that it is critical to be on the same page with each other

## 7- HONOR IMPORTANT RELATIONSHIPS

when dealing with finances. I am nothing without my better half, which is my wife.

Your friends can also be a great networking circle for you. As I told you earlier in this book, as a college football player, I did not think the coaches treated me fairly my first year at Western Carolina University. Although I did not think I was being treated fairly, I still displayed kindness and friendship towards them when I talked with them. My parents taught me to treat others nicely even when I may think that they were not treating me fairly. Love can win people over and help you build a network. The coaches at Western Carolina networked for me and helped me get my first job when I graduated. If I had displayed anger towards them, I am not sure they would have said good things about me to make sure I was hired by this major manufacturing company. I am very fortunate to have a network of friends today. I know they have my best interest in mind, and I do the same for them.

## LESSON:

**Always work to build a network of friends that you can count on to be there when you may need a little help. This network of friends can help you on the journey towards success, so make sure you surround yourself with people who you can share your dreams and goals with and vice versa. Get in the habit of opening doors for others and they will do the same for you.**

**HOW TO BEAT THE ODDS-BE DRIVEN!**

---

*We need others around us so make sure that you encourage others and let them know when they do something that deserves a pat on the back. People will forever remember how you made them feel.*

*- Eddie Maddox, Ph.D.*

---

# CHAPTER 8:
## *Make God Your Foundation*

When I was a freshman in high school, there was a group called the Fellowship of Christian Athletes that met monthly. It was during one of those monthly meetings when I really accepted God as the head of my life. Throughout high school, I stayed very active in the Fellowship of Christian Athletes. Even in college, I made sure I attended church and also supported the local church. I let my faith continue to grow and even today, my wife and I are avid and active participants at a local church. God is the creator of everything around us. God has a plan for all of us. It is important to understand the plan that God has in store for you. We are all special and God has given us all special talents and gifts. Speak with God on a regular basis and you will see the plan that God has for you. A great friend of mine always talks with me about how I live out many scriptures in the Bible. In this chapter, I will share important scriptures by which I live. Each of these verses manifest some characteristics I live by on a daily basis. I will discuss two of my favorite verses first, and then I will share other verses that I find important.

# HOW TO BEAT THE ODDS-BE DRIVEN!

**Proverbs 18:16 NIV**

*A gift opens the way and ushers the giver into the presence of the great.*

This is one of my favorite scriptures that has been a guide for me over the years. I've seen this scripture play out through my life, especially in the abilities I had in the classroom and on the football field. My football skills allowed me to obtain an undergraduate degree and become the first (generation) in the Maddox family to graduate from college. My education has generated substantial financial gains for me and my family. Football also taught me to work extra hard, so it built a work ethic inside me that is unwavering. I believe in doing things right and not taking shortcuts to accomplish tasks. I also hold others around me accountable to this same approach. Obtaining my advanced degrees was a goal I set for myself when I was in high school. While working on my Ph.D., a professor recommended that I also become a professor once I received my doctorate degree. I did not really give it much thought, but once I received my Ph.D., a local college called to ask if I would be interested in teaching classes in the evenings. I was not interested at all in teaching, but the school's dean of the business school was persistent and would not take "no" for an answer, so I finally agreed to teach a class.

## 8- MAKE GOD YOUR FOUNDATION

After the first night, I had a major change of heart about teaching. I enjoyed that first night and started teaching every term for this school. This is a Christian university and they aligned with my beliefs and values. I was doing something that I really enjoyed and my new-found enjoyment provided a second stream of income. This became very rewarding and really changed the game financially for my wife and me. I accidentally stumbled into a second stream of income that allowed us to make a major leap financially. It also allowed us to create a debt elimination plan that was very aggressive all while doing something I enjoyed.

God cares for all of us, and he will lead you to where you need to be. I did not think teaching at the college level was an area that interested me. God made sure others were persistent to persuade me that teaching was a good opportunity for me to pursue. I have my daily talk with God through prayer, and I can assure you that God will never leave you alone. He will help guide you through the wilderness on the journey towards success. God has all powers and He can help you get through challenges and obstacles. God has always been present in my life. It is God and his mighty works that have allowed me to fight through challenges. Each of us will always face challenges. The key is to never give up and be determined to fight through the challenges with the support of almighty God. God wants us to have determination.

# HOW TO BEAT THE ODDS-BE DRIVEN!

Determination is what happens when someone will never give up and continues to fight even when the odds may be against them. God's power equals determination. As a little boy, I learned that determination was the approach to take in order to overcome challenges. Determination will lead you to a destination. The destination that I was focused on achieving was success. You can achieve a high level of success with the help of God guiding you through the journey and by being determined to make it through whatever situations have been placed before you. I have been placed in many challenging situations while going to school, playing football, working on a job, and functioning in our society. The key is to recognize the situation, create a plan to get through the situation, and be determined to get through it.

Remember that challenges do not last forever. Challenges will stress and test you and your faith. When you fall down, you must get back up again and again. Do not stay down because failure should be used as a learning tool so you do not repeat the same mistake. Allow God to order your steps so you learn from all situations, whether it is a success or a failure. My parents taught me to have a strong faith in God and they also taught me to be faithful to God. Having faith in God and being faithful to God will allow you to receive God's favor. Additionally, I learned how to sow seeds through God's direction, for we truly reap what we sow.

## 8- MAKE GOD YOUR FOUNDATION

**Genesis 8:22 NIV**

*As long as the earth endures, seedtime and harvest, cold and heat, summer and winter, day and night will never cease.*

The principle of seedtime and harvest refers to sowing and reaping. We reap what we sow. My goal is to sow seeds and water these seeds to see them grow. It is important for me to encourage others and work with them in pursuit of their personal goals. I also try to help others financially at times by offering advice or helping them monetarily. I treat people with respect and dignity regardless of their status in life and regardless of their religious beliefs, background, or ethnicity.

Over the years, I have worked to make sure that I sow seeds into young people. The youth represent the future of our world. It is necessary for us to show them how to navigate the journey of life. Spending time with others and showing them things that we have learned is critical so that we are developing wisdom for future generations.

The youth will help make our world more technologically advanced. As they build this higher level of technology, it is necessary for people like me to share information that can help them understand why things are the way they are, how to best perform certain duties, and how to cultivate future harvests. I want them to become sowers of positive seeds themselves.

## HOW TO BEAT THE ODDS-BE DRIVEN!

I coached Little League football, basketball, baseball, and track for many years. It was a blessing to develop young people with talent in all these sports, but the thing I enjoyed more than anything was teaching young people discipline, determination, and drive. It is a wonderful feeling to see many of these young people now married with children as successful adults. This tells me that the process of sowing seeds works when they are watered and cultivated. These seeds turn into a great harvest that will generate a greater harvest because they will continue to sow seeds. God gives seed to the sower.

## *Fatherhood*

A great friend of mine observes that in the Bible there are multiple passages that instruct us on the type of fathers that we should be. As a father has compassion on his children, so the Lord has compassion on those who fear him (Psalms 103:13 NIV). As fathers, we are instructed to love, instruct, guide, warn, train, discipline, nourish, and supply our children's needs. My friend has joked with me and said that if I had been his father he would probably be the president of the United States. The father of a righteous child has great joy; a man who fathers a wise son rejoices in him (Proverbs 23:24 NIV). My goal is to offer fatherly advice to everyone that I come in contact with because I want to see a world that is full of love and success.

## 8- MAKE GOD YOUR FOUNDATION

Because the lord disciplines those he loves, as a father the son he delights in (Proverbs 3:12 NIV). My wife and I have worked extensively with our two children sharing God's plans for them by raising them in an environment in which God is first. We taught them God's principles. Fathers, do not exasperate your children; instead, bring them up in the training and instruction of the Lord (Ephesians 6:4 NIV). Building a strong foundation with your children is key to becoming a great father. Being a father is one of the greatest feelings in the world. It is something I took seriously, and I made sure my children understood my goal was to guide them from being dependent children to independent adults. Endure hardship as discipline; God is treating you as his children. For what children are not disciplined by their father (Hebrews 12:7 NIV)? It really makes me feel good to see my children as adults and see how they carry themselves based on God's principles. God has provided me with wisdom and insight; therefore, I must use this to help others become successful.

**Proverbs 11:14 NIV**

*For lack of guidance a nation falls, but victory is won through many advisers.*

## HOW TO BEAT THE ODDS-BE DRIVEN!

In the multitude of counsel there is safety. So, while God has blessed me with wisdom and gifts, I am humble enough to take into account the perspective of others. This is where friends can help keep you headed in the right direction. When situations arise that require feedback from friends, I know I can always count on them to give unbiased advice.

There are other verses in the Bible that can be used by individuals and businesses to help motivate themselves and others to work towards achievement and success. Here are a few I especially recommend:

## *Wealth and Globalization*

**Deuteronomy 8:18 NIV**

***But Remember the Lord your God, for it is he who gives you the ability to produce wealth, and so confirms his covenant, which he swore to your ancestors, as it is today.***

God has given each of us special talents and he expects us to use these talents in business to create a society that is in harmony with the environment. God gives us these abilities and special talents to produce wealth. This means that we must spread this wealth throughout all nations. Globalization is the avenue that can allow us to help spread this wealth. God has allowed people before us to help start this process of globalization and we must continue this process to make the world a better place for all life.

In the early years of manufacturing in this country, the majority of items were made in-state, but eventually, manufacturers realized that they could go outside of their state lines to have goods and services produced. Eventually, we realized that we could get products and services from other countries more economically than in the United States. This is the globalization process God has allowed to take place. It is critical to understand our global environment can offer many opportunities for you to be successful. The world is our neighborhood, and due to our technological breakthroughs, we can communicate and work with someone in India or other parts of the world instantly. This global approach can be used on your journey towards success.

## *Innovation*

**Romans 12:2 NIV**

*Do not conform to the pattern of this world, but be transformed by the renewing of your mind. Then you will be able to test and approve what God's will is–his good, pleasing, and perfect will.*

# HOW TO BEAT THE ODDS-BE DRIVEN!

We are commanded by the Bible to be innovative and we should carry this innovation over to our businesses. Each of us is unique and we should use these unique qualities in our businesses to be both specialists and generalists. As a specialist, you will be an expert in some area such as engineering, quality, or other management roles. As a generalist, you will know about a variety of business disciplines and you will use this knowledge to create innovative strategies to build a strong company. Our society must always strive to make improvements. By renewing ourselves through learning new information, we can keep the process of continuous improvement in place.

It is good for every individual to renew him or herself. By acquiring knowledge, you can help implement new things. I am convinced that everyone has a special talent. You may be that person who can help find the cure for a terrible disease that may be tearing through the world. Dr. Martin Luther King, Jr. said as long as disease is around it will reduce the length of life. It is critical that we allow every person to reach their potential because that may help prolong the length of time humans live by eliminating diseases. We need people of all walks to be focused and work to be the best that they can be in their fields, and this will lead to cures for diseases and allow us to make other major contributions in our society. We will be disappointed sometimes as we work to be the best. Dr. King said,

## 8- MAKE GOD YOUR FOUNDATION

"We must accept finite disappointment, but never lose infinite hope." By having and displaying hope in your life, you can help yourself on the journey towards success because it gives you focus and purpose. It is necessary to dream and drive towards innovation. Being innovative is a pathway to success and this quality has been observed in successful people. We can accomplish things when we never give up and stay focused on the task that must be completed.

## *Prayer*

**I Thessalonians 5:16-19 NIV**

***Rejoice always, pray continually, give thanks in all circumstances; for this is God's will for you in Christ Jesus.***

It is important to pray and give thanks to God for the blessings he gives you. Every day will not go the way you plan, but it is important to remain faithful and prayerful through these tough times as well as in good times. It is necessary to thank God all the time. We must thank God when we develop a new technology, and we must thank God when our processes are not working properly. God knows the big picture, and he allows things not to go our way sometimes to make us stronger. He also leads us through tough times to prepare us to make it through other journeys. Always give God thanks because he is our protector.

# HOW TO BEAT THE ODDS-BE DRIVEN!

I have asked God to bring me through many situations. One such situation God guided me through was my college football experience. Many star high school athletes give up once they get to the college level due to the challenges that they face. A high school star athlete is accustomed to everyone considering them as a special talent and person. When you are a star high school athlete, you are used to everyone praising you for your talent. While you were "the man," or the top athlete in high school, once you enter the college level all this changes. All the players on the team are considered special players because of the great ability that they displayed in high school. The top talented high school athletes are the ones that get college scholarships. The playing field becomes a battlefield filled with players trying to gain a starting position so they can shine on Saturdays. Many kids get lost in the shuffle of all the talent on the team and do not develop and become the best that they can be academically and athletically in college. Fortunately, God guided me through academics and athletics through prayer. God helped me through the situations which were beyond my control and his favor allowed me to stay the course. Prayer is a daily part of my life. I have prayer meetings with God continuously throughout every day and at night.

# 8- MAKE GOD YOUR FOUNDATION

Every year for the past ten years, I have participated in a fast at the start of the year. A great friend of mine introduced me to this fast and it has helped me become more focused in all the things that I set out to accomplish. My friend introduced me to the Daniel fast, which is based on the dietary restrictions maintained by Daniel in the Old Testament (Daniel Chapter 10). During this fast, I give up certain pleasures for twenty-one days. I give up sweets, meats, and bread. By giving up these items, I must use willpower to maintain discipline throughout the three weeks. Over this twenty-one day period, I have constant dialogue with God through prayer. By day twelve, it usually becomes really hard, but I pray and ask God to give me the strength to overcome this situation of craving the things I want. The fasting process has given me a strong prayer relationship with God. God has shown me through this process that I can accomplish my goals if I stay committed and never give up even when it feels like I cannot go on any longer. Since starting this fast, I have enjoyed the journey towards success even more. I consider it a journey because success is something you should always strive to reach no matter what your status is in life. It is a never-ending journey and requires great focus and determination to stay the course.

This process has taught me to mold my environment and not become a product of the environment. As I look at magnificent leaders throughout history, it is clear they were molders of their environments. Mahatma Gandhi was able to influence and motivate others to get involved in bringing Indian independence. He was determined to excel at whatever he did. He effectively used the policy of non-violence, protest, and civil disobedience to successfully lead India to freedom from colonial rule in 1947. Dr. King used this same non-violence, protest, and civil disobedience policy to lead a movement that required him to be strong through situations in which he was treated harshly. He had a goal to achieve, and he achieved it by loving others even when they showed hate towards him. His work has really opened doors for equality and fair treatment, and it allowed him to have an opportunity to be successful. Similarly, Nelson Mandela was elected president of South Africa from 1994 to 1999. He was an anti-apartheid revolutionary who focused on dismantling apartheid in South Africa by working to eliminate institutionalized racism and advancing racial reconciliation. Mandela was fearless and determined with unyielding perseverance, and he was devoted to achieving his goals. I can name many more great leaders, but it is obvious that these leaders molded the environments around them. A true leader molds the environment and builds a culture that continues to work towards the goals that he or she has set.

## 8- MAKE GOD YOUR FOUNDATION

## *Pride*

**Proverbs 16:18 NIV**

***Pride goes before destruction, a haughty spirit before a fall.***

I have seen many people struggle with pride. As individuals, we do not know everything, and it is okay to tell others you do not have the answer to a particular problem or situation. Make sure you are truthful in all situations. Don't allow your pride to get in the way of you making the best decision in your personal life or for your business. Pride can destroy you and your business. People who do not wear their pride on their sleeves will be more successful in all areas of their lives.

These types of people are more likely to ask for help, seek wisdom from others, delegate projects, and get second and third opinions to review their work. Proverbs 16:18 states, "Pride goes before destruction, and a haughty spirit before a fall." You are less likely to fall when you have accountability and after you have addressed any areas of pride. We are one society and we are connected with others. Your success can help me and others become successful. Allow others to give you ideas and thoughts that may go against something you've done in the past or the way you have performed a certain task. When you notice your own resistance to take ideas and learn from others, that is a clear indication that you have let your pride take over. Successful people are able to set their pride aside so it will not become an obstacle to their success.

## *Planning and Preparation*

**2 Corinthians 9:8 NIV**

*And God is able to bless you abundantly, so that in all things at all times, having all that you need, you will abound in every good work.*

God does his part when we do ours. He wants each of us to do our best and our best will yield the best possible results for our lives. In order to get the best results, we must be willing to plan and prepare in advance. The planning and preparation process allow us to create a system that will help us make decisions about goals and activities. The planning process includes God, family, and friends. God will help us through this process and provide all the things that are needed to formulate and carry out plans. He will bless you at all times if you trust in Him.

I encourage everyone to define a plan for things you want to accomplish. If you want to take your family to Disney World, then you need to start with a plan. How do you think a trip to Disney World would turn out if in the middle of the school week you come home and tell your wife and kids, "We are going to Disney World right now, and everyone needs to be in the car in two minutes?" This trip would not turn out very well logistically and probably financially either. A good trip to Disney World starts with a plan. You will plan the trip months to years in advance. This will allow you to define a budget for the trip, your travel dates, the length of your vacation, and your lodging plans, just to name a few key logistics. You will more than likely enjoy this well-planned trip. This type of planning is critical on the journey towards success.

## LESSON:

**I hope you can see that "success" isn't just a destination. You can experience success each day in every endeavor that you properly plan. You can plan a successful vacation to Disney World, a successful meeting at work, a positive financial portfolio, a successful marriage, and so on. Think of "success" as constant pit stops on the journey toward your ultimate destiny and promised place.**

A person building a house creates a plan before they start building the house. It is necessary to understand the critical parts of a plan. In the case of a house, the critical part starts with creating a blueprint for the design of the house and laying the foundation. A home not built on a solid foundation like the one in the Bible (Matthew 7:24-27) that is built on sand will sink or erode away very quickly. Hence, start out with a plan to make sure you are headed in the direction you must go to achieve your goal.

# HOW TO BEAT THE ODDS-BE DRIVEN!

## *Leadership - lead well.*

**1 Timothy 3:4 NIV**

***He must manage his own family well and see that his children obey him, and he must do so in a manner worthy of full respect.***

Although this scripture refers to family leadership, its principle should be applied to our professional lives as well. It is very important to have good leaders in place in your business. Good leaders have the ability to motivate or stimulate people to utilize their potential. It is necessary to have good communication skills so you can work with people daily to guide and inspire them to achieve the goals that have been defined for the organization. It is also important for the leader to have advisers. Advisers can provide advice based on their education and experiences that can help you foresee issues before they occur and become problems. This allows you to lead proactively. As a leader, select a cross-functional group to advise you and this will help prevent your business from failing.

Dr. Martin Luther King, Jr. was a great leader. He and other leaders that I have discussed have those special skills that allowed them to be great communicators. Over the years, I've noticed other characteristic in great leaders, including:

- They stay positive through all situations even when the situation may be horrible.

- They have confidence and they show this confidence as they are leading. Their confidence impacts the group they are leading in a positive manner.

- They have a sense of humor. They are able to connect with people and share humorous stories to help others see how goals can be accomplished even during tough times.

- They embrace failure and setbacks. They will not blame others for their failures, but instead will step up and take ownership in situations even when someone else may have caused the failure.

- They will not allow fear to take over and cause pain for the group.

- They are willing to take feedback. (Feedback is not always positive, so as a leader, you must be able to handle criticism. You are not perfect, and when you make mistakes it is okay to listen to what others have to say and not chastise them for pointing out something that may have gone wrong.)

- They listen to what others have to say.

- They are inspirational. They inspire others to accomplish tasks. On any journey towards success, you will need others on your team and it is necessary to be able to inspire others.

- One final point that I have witnessed in great leaders is they use past situations as lessons learned and they make sure that the past is used to help them make better decisions in the future.

# HOW TO BEAT THE ODDS-BE DRIVEN!

My parents, teachers, coaches, and friends helped me develop leadership skills that have helped me on my journey toward success.

Whenever I go through a situation, I always make sure I discover the lesson and understand the blessing from it. It is necessary for us to understand how and why a situation may have occurred so that we can repeat it if it was successful and make changes to it if it was not successful. Always be thankful to God for teaching you a lesson or allowing a situation to be a success.

## *Trusting God*

**Proverbs 3:5-6 (NIV)**

***Trust in the Lord with all your heart and lean not on your own understanding, in all your ways submit to him, and he will make your paths straight.***

You can be successful in life. The key is to trust God and He will lead you in the correct direction. Even when the situation may seem too difficult, put your faith and trust in God. Success comes to those who refuse to give up no matter how hard the task at hand may appear. It is necessary to keep your eyes on your goals, and you will reap success.

# CHAPTER 9:
## *Learn from Coaches and Teachers*

***Train your mind so that you can build a solid foundation which will support your worth.***

My parents were my first teachers. They both encouraged me to be the best at whatever I set out to accomplish. They taught me the importance of school. My parents also taught me the importance of independence. I was in the seventh grade and wanted a moped. This is a motor bike with an engine, but it doesn't have the speed or power of a full-fledged motorcycle. My parents told me I could get a moped, but I would have to pay some of the costs. They paid the down payment on this moped, but I was responsible for making the twenty-five dollar per month payment to them. This was the first time I was introduced to being responsible for a debt. At that age, I knew that having to pay a bill for something every month was not fun. I used money made during a summer job to make the monthly payments. Eventually, I saved enough money to pay off the moped and

## HOW TO BEAT THE ODDS-BE DRIVEN!

eliminated the monthly debt. This was a critical learning point for me in understanding what debt was all about and I realized I did not want to have debt.

I have always enjoyed talking with older people. When I say older people, I mean people that are seventy years old or more. As a little boy, it always fascinated me to listen to information that elders would share. I was a sponge that absorbed the information. One of the best lessons I received from an older person was to make sure that I always listened to my parents. They would stress to me that my parents wanted to see me become successful by telling me things that they had learned over the years. This wisdom from my elder is still very true. My parents shared details with me about school and how to function in this world. This information has helped me stay on the right roads and headed in the right direction. The older adults also shared with me that I should always respect my parents because if you will respect your parents, then you can live by the rules in society and not disrespect authority figures (such as teachers). I have watched many people over the years who were very disrespectful to their parents and this disrespectful mentality transferred to other parts of their lives. These individuals refused to follow the rules of society and the authority of others. These individuals have not lived successful lives but have struggled by losing purpose and direction in their lives along with breaking laws that placed them in jail and

## 9- LEARN FROM COACHES AND TEACHERS

crippled their chances for success. It is very simple: if you respect your parents, then you will have respect for others and the rules of society.

Older people have also discussed with me that those who sit down will go down. In other words, they were telling me to stay busy. How many times have you read about a person that worked for a company for forty years and retired? And then after they retired the person became ill and could not overcome an illness. One older person had seen this happen over and over time after time. In your early years, it is important and critical to find something that you enjoy and that will keep you busy so that you do not go down physically and mentally. It is clearly wisdom that the older people were sharing with me. Hopefully, one day someone will consider the information that I am sharing with you to be considered "wisdom" as it pertains to success.

## *Good Teachers*

As stated earlier in the acknowledgements section, as a young student, I was fortunate to have wonderful teachers in Head Start, elementary, and middle school. My good fortune continued through high school and college. I would like to thank all my teachers because each one helped create and mold me into an individual who always strives for more knowledge.

# HOW TO BEAT THE ODDS-BE DRIVEN!

My Head Start teacher, the late Mrs. Elsie Thompson, created a great foundation in me. She taught me the importance of being a role model student and the necessity of changing society for the best by learning new things and being involved.

I want to give two of my elementary school teachers special thanks. Mrs. West and Mrs. Pat Evans were excellent teachers in those early years of my education experience. Mrs. West really gave me the confidence to be a great math student. She would always call on me to come to the board to work out math problems in front of the class. This gave me the confidence to know that I understood the material well enough that the teacher wanted me to show the rest of the class. Similarly, Mrs. Evans helped develop me into a student who always wanted to be better at my school work. When I turned in a less-than-perfect assignment to her, I would stay in at recess time on my own accord and re-work the homework to make sure I completed everything correctly. They both taught me the key skills to being a successful student and encouraged me that there would be a payoff to all my hard work.

In middle school, I met a physical education teacher who gave me purpose and meaning. Coach Shabazz taught me the value of physical fitness and the importance of taking care of the mind. He gave me the courage to always challenge myself to be the very best.

## 9- LEARN FROM COACHES AND TEACHERS

He was an outstanding role model. He helped me understand that I must always take care of my body. This includes both physically and mentally. He was full of life lessons. Through his teachings, it was clear to me I did not need to put alcohol or drugs into my body. I made a commitment to myself at an early age based on things that Coach Shabazz taught me to never put alcohol or drugs in my body, and this is a commitment I continue to uphold even today. He helped me to understand how important it was to run track which helped develop my football speed. It is clear to me that once I started running track, my speed as a football player became faster and I was scoring on long touchdown runs. The lessons I learned about health, fitness, mental and physical agility from Coach Shabazz created a solid foundation for me in track, football, and life in general.

In middle school, Mrs. Williams taught me the importance of utilizing my talents in mathematics. I consider her not only a math teacher, but a counselor who made sure I enrolled in the correct classes as I moved on to high school. Mrs. Williams's concern for making sure I enrolled in the right courses in high school was the greatest gift any teacher could have given me. I also had a terrific teacher in high school. She was my math teacher, Mrs. Emma Doughty. She would not allow me to doubt myself. Thanks to all these outstanding teachers, I have been able to achieve my goals.

# HOW TO BEAT THE ODDS-BE DRIVEN!

My high school football coach, Preston Cox, helped me develop into an outstanding football player. He made me a starter as a freshman. This was a huge confidence booster for me because it made me realize I had talents in football that could help build my future. He was always there as a coach and counselor to make sure my abilities were being utilized so I became a better leader and team player. He worked with me through injuries and exceptional times when I was winning awards. All of these individuals helped mold and shape me into the man that I am today. They are all responsible for the level of success that I have achieved. When I think about all of them, it inspires me to give my time to young men and women to help them build a sound foundation in their lives that will lead them to a life of success. Having people in your corner is necessary to make sure you can successfully navigate the long journey of building a successful life. I am very fortunate all of these people invested in me because they wanted to see a young African American male grow up to be a successful man. I am truly grateful for them believing in me, the little shy boy from Pendleton, South Carolina. Making these type investments in others will help build a society of successful people.

*The obstacles that you face will not be easy to overcome but they will teach you how to navigate on a rough sea so that you can reach your goals.*

- Eddie Maddox, Ph.D.

# HOW TO BEAT THE ODDS-BE DRIVEN!

# CHAPTER 10:
## *Set Clear Goals and Measure Progress*

Goal setting consists of understanding what you want to accomplish. Establish a timeframe to complete the goal and define a measurable objective to determine if the defined goal has been achieved. Goals should be used for all facets of your life, but I highly recommend that you set goals for achieving your financial freedom. People will focus on things that are important. Your financial status is definitely an important part of your life. Financial success should include setting financial goals. Examples of financial goals are as follows:

- Save $50,000 within two years from June 23, 2020 and invest it in stocks.
- Eliminate all debts in five years from June 23, 2020.
- Save and invest twenty percent of my income each month.
- Create an emergency fund that consists of six months of my yearly income by June 23, 2022.

# HOW TO BEAT THE ODDS-BE DRIVEN!

It is critical to set reasonable goals that you can attain. Once the goals are set, it is necessary to measure where you are at a set frequency. Measuring the goals will create accountability for you. You now know and understand that you have a target to meet.

You and your family should share similar goals so that everyone is headed in the same direction. Goal setting must be clear, in place, and implemented by the husband and wife for the household. As I mentioned earlier, planning is the key to everything you do because if there is no plan or blueprint to define how to achieve your goals, then achievement will be at a low level or will not exist at all. Adequate preparation leads to a smoother implementation process for the household. Goal setting for the household must be clear and implemented by both husband and wife. A household in which the husband and wife are not moving towards the same goals will not be as financially successful as it could be. I consider commitment between the husband and wife to their joint goals as the most critical factor in implementing and sustaining a budget that will help a family reach financial security.

You and your spouse should set goals that follow the SMART approach. The goals should be specific, measurable, attainable, realistic, and time bound. You should measure how you are performing

## 10- SET CLEAR GOALS

against the goals that you have set in a specific area. Here is an example of a goal. You and your spouse will payoff the $20,000 student loan by September 5, 2025. This goal defines that you will pay this off in five years from September 5, 2020. This means that you will have to pay the principal balance on the loan down by $4,000 every year. Now you must create a plan of how you will achieve this each month.

If you are currently in debt, then you need to create a plan to eliminate the debt. If you are not in debt, please be intentional to not get into any debt. I call it being content with what you have until you can afford to pay for what you want. I totally understand that most people will have to go into debt to buy a house, so create a plan as to how you will pay the house off earlier than the loan terms. Create an environment in which there is a partnership and relationship that keeps communication open and makes sure that you and your spouse are headed in the same direction.

As I told you before, I graduated high school as number six in my class of 210 students. My teachers in the early years gave me a solid foundation to build upon. My favorite subjects in high school were math and science. I really enjoyed figuring through mathematical problems in algebra and calculus. It was also fun working through science experiments and building electrical systems in these classes. This was the foundation that I learned early in my educati-

## HOW TO BEAT THE ODDS-BE DRIVEN!

onal journey and I thrived at working through these areas in school. My next step was to build on these skills in college. Based on my liking for math and science, it was clear that engineering would be my choice of study. I started out in liberal arts because of the college that gave me a football scholarship but had the hopes of transferring to a technical school. I was able to achieve my goal and received a degree in manufacturing engineering.

After graduating, I went to work for a major automotive manufacturer that supplied steering components, transmission components, and bearings to various automotive manufacturers, including Ford, General Motors, and Chrysler. It was very exciting to actually be working as an engineer to build cars and trucks. My manufacturing career spans more than thirty years. I have held many different positions throughout my career in engineering and management. The different positions I have held include project engineer, application engineer, product engineer, engineering manager, operations manager, and plant quality manager.

After I had been working for twenty years, my kids had become of age and this coincided with the inspiration from my wife to go back to school to get my master's degree in business administration so I had no excuses to enroll. I enrolled in evening classes and obtained a master's of business administration (MBA). I realized after I received my master's degree how much I missed going to school. At that time, I decided to continue with my education, and I enrolled in a doctoral program. Obtaining my doctorate was on my list of priorities that I planned to accomplish. I obtained my Ph.D. in organization and management. This was a milestone for me because it gave me a major sense of accomplishment. Obtaining my Ph.D. was one goal I could mark off my bucket list.

I'm not through yet. Each year I set goals and create a strategy to help me achieve them in both my personal and professional life. I review the goals and strategy monthly and change it as necessary to make sure that I continue on the correct path. You can be successful in anything that you choose to do. In order to obtain the level of success that you want, you must set your sights on the goal and do not allow anything to break your stride. There will be many things that will try to hinder you in reaching the goal but staying focused will help you realize your DREAM.

Plan to give your best effort. When things may appear to be tough, this is the time to give an all-out effort so that you can finish strong. Through life you will come upon situations that appear to be very hard. I consider these the speed bumps of life. It is necessary to

## HOW TO BEAT THE ODDS-BE DRIVEN!

look within yourself and maneuver over these speed bumps because once you go over the speed bumps, you will be able to taste success.

*I want the obstacles that I have dealt with and have overcome to be a beacon of light and a strategy to help others become successful.*

*- Eddie Maddox, Ph.D.*

# HOW TO BEAT THE ODDS-BE DRIVEN!

# CHAPTER 11:
*Drive to Achieve*

In order for you to achieve at a high level, you must have a drive inside that is highly competitive and you must be motivated to win. My drive is to accomplish the task at hand through the proper means. This means that my journey to succeed is strong, but I do not compromise it to involve illegal, immoral, or unethical acts.

The journey to success must be true and it must be focused. If I am taking a test and the maximum possible score is 100 points, then I will do everything possible to achieve that score, as long as it is legal, ethical, and moral. This means that I am going to start studying for the test on the first day of class. You cannot wait until the night before the test to start studying. You must have a drive inside that forces you to create a study guide after the first class meeting and begin to study immediately. I used an academic analogy to explain this principle, but the same type of drive must be in place for everything in your life. Have a strong drive when completing a work project, a home project, or even when you're doing something

## HOW TO BEAT THE ODDS-BE DRIVEN!

good for others. This should be your mode of operation. This mentality should carry over into everything that you do and you will see that it is a key ingredient to being successful. Make sure that you have DRIVE. Here is a way to think about and remember "drive."

- **D is for desire.** Desire means you want to complete a task because there will be enjoyment or satisfaction once you do.

- **R is for readiness.** This means that you will be prepared and prompt to carry out tasks.

- **I is for independence.** Independence means you have freedom and autonomy to complete tasks. Individuals who are tied to debt or other financial worries are not independent. I have found people who are independent will speak up and challenge the status quo because they are not worried if someone will give them a bad review for their job performance.

- **V is for victory.** Victory means you come out successful even if you are in a highly competitive situation. You have that fire inside of you that presses you to be a master of achievement.

- **E is for enthusiasm.** Be enthusiastic about completing the goals and winning the challenges.

## 11- DRIVE TO ACHIEVE

### *Finish Strong!*

Sometimes the workload of school, work, and home life gets very heavy. God will help you through these times. Continue to strive for the goals you want to achieve. With God on your team, you can overcome any obstacle. Remember, continue to strive, because God is on your side. He will get you through the tough times and remain with you through the good times. Always put God first, and He will lead you in the correct direction. He will help you over those speed bumps in life, and He will help you finish strong. You can obtain success by allowing God to order your steps. Success requires you to give your best effort in whatever you are doing. If you do not try to complete a task, then it will not be completed. You cannot receive a college degree if you do not complete the work. You cannot become an Olympic champion if you do not train. On the other hand, you can be successful at anything you want if you meet the required work for a task. A solid example that I experienced recently on finishing strong was when my son and I were jogging. We were out on a three mile jog and he was pushing the pace of the jog to a fast pace. It was taking a physical toll on me and he was working hard as well. We were physically exhausted towards the end of the jog. My son's motto is to finish strong. Of course, I want to always finish strong as well. Although it appeared that we did not have much energy left as we were closing in on the last 400 yards of the jog, we both made a point to increase our speed or pace so we would finish strong. Success requires work. Good luck as you take the next steps on your journey towards success.

## *Achieve Prosperity*

**Joshua 1:7-9 (NIV)**

*Be strong and very courageous. Be careful to obey all the law my servant Moses gave you; do not turn from it to the right or to the left, that you may be successful wherever you go. Do not let this Book of the Law depart from your mouth; meditate on it day and night, so that you may be careful to do everything written in it. Then you will be prosperous and successful. Have I not commanded you? Be strong and courageous. Do not be terrified; do not be discouraged, for the LORD your God will be with you wherever you go.*

Continue to work hard to complete your mission. God has a plan for everyone, and God will use your strengths and weaknesses to carry out the plan that is in store for you. Continue to allow God to guide you, and you will reap success! Prosperity can be achieved by those who will be strong and courageous to obtain their goals. Successful people write down their goals and review them on a regular basis. I create goals every year prior to my Daniel fast. As I am fasting, I pray over these goals multiple times per day. Once I am done with the fast, I review the goals regularly and measure how I am performing. Your goals can help you on this journey to success. I have found that you can set many small goals that can add up to a large goal that leads you to being successful.

## *Build Trust*

To be considered a true leader, you must build a trusting environment. A trusting environment is built by being a great communicator. The key is to communicate clearly, concisely, and honestly. People like to work with others who do what they say and say what they do. In other words, people like to work with others who are reliable, dependable, and honest. You must be authentic. You also must build commitment amongst others. People who are committed will make sure that the game plan is carried out. I have found in a household where there is communication and commitment the couple will flourish if they are working toward common goals.

Building trust should be something that consumes you and you must exemplify a trusting environment in all your dealings. When you have a high degree of confidence in the words and actions of another person, it means you trust that person. There are two forms of trust to which we are most often exposed: trust based on threats and deterrents and trust based on the power of personal relationships.

# HOW TO BEAT THE ODDS-BE DRIVEN!

The approach of using threats and deterrents to gain trust works for many people. In this form of trust, a person cooperates as promised out of fear of being punished. An example of this type of trust is when you work the correct number of hours on your job each day because you are afraid that you may lose your job if you do not work as required. Another example is when you perform work and trust that you will receive a paycheck for the time you worked. This type of trust is associated with transactional contracts. The second form of trust is associated with the power of personal relationships. This type of trust means that you understand a person's wants and desires. You have a very strong and positive relationship with another person and you are willing to allow this person to act on your behalf. An example of this type of trust is when you will allow your spouse to discuss your medical condition with the doctor. You know that your spouse will be working on your behalf to make sure you are well taken care of by the doctor. This type of trust is based on the following factors:

1. Familiarity
2. Shared experiences
3. Reciprocal disclosure
4. Promises
5. Demonstrations of not exploiting others and their vulnerability

**Familiarity** is when a person trusts another based on a relationship that they may have or had in the past. **Shared experiences** refers to the times that you spent with this person doing a similar activity. You can have shared work experiences, military experiences, and college experiences with others. **Reciprocal disclosure** is when you and another person have shared information about one another that makes your interpersonal relationship stronger and brings it closer. **Fulfilled promises** are promises that another person has fulfilled, and a strong history has been established that shows if he or she makes a promise, this person will meet it. **Demonstrations of not exploiting others and their vulnerability** is when a person has an opportunity to exploit another person, but did not take advantage of them even though that person was in a vulnerable position. This type of trust can help you build a strong environment that allows you to get things accomplished because when others trust you, they will work to make sure actions are carried out.

## *Be Visible to Your Family, Friends, and Co-workers*

Successful people are visible people. You should be visible to your family, friends, and other associates. By being visible, they know that you care and that you are putting in the effort with them as you are on the journey to success. What can success do for you when you

## HOW TO BEAT THE ODDS-BE DRIVEN!

have no family or friends to share it with? The point that I am making is do not get caught up in success and fail to spend time with your family and friends. I have seen individuals in the past work so much that they do not get a chance to see their kids grow up. Family must come before money. Your family will be there for you when obstacles arise. You need to take time out to spend with family and friends so that they can share in the success once you reach it. Success is no fun alone, but it is fun when you have family and friends to share it with. I mentioned creating a second stream of income. It is good to create a second stream of income, but make sure that this second stream of income does not get in the way of you spending time with family and friends.

Family, friends, and co-workers can help you be successful on your journey. Always treat others how you want to be treated. It is important that you set an example for your family, friends, and co-workers for how to be successful. It is a great experience to see others in your life who you've influenced, become successful. Show others what works and will help them to reach success. These same people will also show you things that will contribute to your success. Our lives will be of a higher quality if we work to help others around us become successful.

## 11- DRIVE TO ACHIEVE

### *Provide Resources to Help Others be Successful*

We have enough space in the world for everyone to be successful. There are many ways to provide resources for others to be successful. I teach on the college level and my goal is to make sure I am providing material to the students they can use to be successful on their current jobs or provide them information that will help them be successful on a future job. I have friends that have monthly meetings on certain topics to help members of the community build a successful life. There are many charitable organizations that we support to help others become successful. When we help others, we help the whole world. A kind act towards someone might be the missing piece they need to help them be successful or help them change their life.

### *Coach and Mentor Others*

A person who acts as a coach will help others become better contributors. You can do this by facilitating communications and fostering teamwork. These two factors are critical to the success of others. People should always work together because this builds teamwork. A team is stronger than an individual. Teams can benefit from coaching. Coaches know how to listen and direct, and this works well for teams. A leadership team must be very good coaches

## HOW TO BEAT THE ODDS-BE DRIVEN!

if they want to become successful. In a corporate environment, a coach works to improve employee performance. A mentor helps obtain career aspirations and overall development goals. Over the years, I have been exposed to very good coaches and mentors in various aspects of my life. In middle school athletics, Coach Shabazz was an excellent coach and mentor. Coach Shabazz helped me improve my performance in the 100-meter and 200-meter races. He coached me through this process by having me conduct training routines each day that were focused on improving my performance. He was also a great mentor to me because he helped me develop into a young man with strong characteristics on how to carry myself as a man. In high school, Coach Cox helped improve my performance as a football player through routines that we practiced each day and during off season training.

If there is someone that you can coach and mentor, please take advantage of that opportunity. It is self-fulfilling and rewarding to help someone improve their performance and develop their overall career. My goal is to coach and mentor people. I do this by having discussions with others on how to become better in a sport or a job. Over the years, I have taken young people under my wings to help them improve their football skills. I have helped young engineers develop into engineering managers. I have helped young leaders develop into directors of operations. I have also helped people see

## 11- DRIVE TO ACHIEVE

and understand the importance of controlling their finances. I do this because I want others to be successful. We must work to build a successful society. Each of the people I have mentored may mentor others and this process creates many mentors and builds a community of successful people. Lastly, make sure that you are coachable. This means that you must listen and follow the instructions that your coach is giving you. Coachable people are successful people.

## *Teamwork – Work with Others to Build Success*

Teamwork is critical in a successful household. It is also critical when working in an organization. You cannot achieve success without working with others. Employees in the same company should always work together and not compete with each other because this becomes counterproductive for the organization as a whole. It is necessary to work as a team and pull in the same direction so others on the team collaborate and embrace the team environment. It is important for every team member to understand how they personally contribute to a team or project. Everyone on a football team cannot be the quarterback. In order for the team to be successful, there must be players willing to play the other ten positions on offense. If this does not happen, then the team has no chance of being successful. All po-

## HOW TO BEAT THE ODDS-BE DRIVEN!

sitions on a team are important and you must have someone carrying out each role so the team has a chance to be successful. The same is true for your personal life. If your spouse is a better money manager than you are, your spouse should control the finances in the household. Teamwork promotes productivity, healthy competition where you are on the same team, and successful outcomes. You and your spouse can be more effective as a team.

It is recommended that teams do the following:

1. Have shared vision and goals
2. Have the skills required
3. Understand respective roles on the team
4. Have strong interpersonal relationship with members
5. Celebrate accomplishments
6. Learn from past failures and successes

One last thought: In order to achieve success, a person needs to be agreeable. You do not need to agree with everything others say. You should carefully consider the direction in which a leader is taking the team, but once a direction is agreed upon, you should support it. If you are part of the team, it is critical to be agreeable to the direction that the team has set even if it may not be your desired approach. I analyze people who are successful, and being agreeable is one trait that I have recognized in all successful people I know.

# 11- DRIVE TO ACHIEVE

## *Be Involved with the Process*

It is necessary for each of us to be involved with many things throughout life. Your family is critical for you to be involved with as you journey through life. Spending time with your family helps strengthen your relations with your spouse and develops your children into responsible adults. A couple must grow, and enjoying quality time with your spouse is a sound way to make sure that you are developing the relationship in a positive manner. Having date night or going to the movies or out to eat is very important for a relationship. I have found that the most critical part of a relationship is being involved and making sure that you and your spouse communicate. Listening to your spouse is also very important. Plan to listen to hear what your spouse is saying and not listening to respond to what she may be saying.

Being involved with your children is also critical. It is our responsibility as parents to develop our children to become productive citizens of our society. As parents we must also listen to our children, and the listening should be to hear what they are saying and not listening to respond. Communication is essential in the development of children. Spending quality time is a method to be involved with your children. Work to help them develop their skills in dealing with others; skills in school; and skills in sports, arts, or other areas that may interest them.

# HOW TO BEAT THE ODDS-BE DRIVEN!

There are similarities to family involvement and job involvement. You also should be involved with processes in your job. If you are part of management in an organization, then you must be involved with the process that is being used to operate the business. Management or personal involvement with your family can make a difference in the level of success obtained by the organization or family. This can help families and CEOs better make decisions on why they should make improvements and see how these improvements will impact the level of success they will achieve. It is very clear that you should work towards continuous improvement and involvement. Continuous improvement motivates one to always challenge themselves to be better. Even if you are successful, you must continue to look for ways to become better, or you may lose the competitive advantage that you have to offer. In an organization, management involvement is the way organizations can motivate all employees to perform at their best. This same motivation works with your family, so be involved in the lives of your children so that they perform at their best. You can only understand the state of the business or household if you are involved on a daily basis. Just as you should help your children daily with their homework, the management team can help employees see the benefit of programs by working closely with them and displaying that this process is a long-term solution to help make everyone's job more effective and efficient. Working with your family in a similar manner will show your children and spouse how to succeed on the journey of life.

During a process, it is important everyone understands that flexibility is necessary so that a company can meet customers' changing needs. You need to have and show flexibility with your family as well. Your involvement is critical to the success of your children and the household. Lack of involvement or commitment has hampered the development of children and families. Lack of involvement with employees in the workplace also hampers the development of employees. I aim to spend personal time with individuals in the workplace to get to know their families and understand what is important to them. I have found when people are treated with dignity and respect, they are more likely to complete their jobs effectively and efficiently.

## *Build Collaborative Relationships*

It is necessary for people to build collaborative relationships with others. When you work collaboratively with others, it allows you to build synergies that create major momentum. Be generous when giving your time, resources, and knowledge. This will allow you to easily build collaborative relationships with others. You need to be intentional when sharing your time, resources, and knowledge, be-

## HOW TO BEAT THE ODDS-BE DRIVEN!

cause these are keys to building a stronger society around us. Collaboration is used to create major synergies for individuals and groups.

I work with teams in a collaborative manner on a regular basis. Many years ago, I worked on a set-up reduction project for manufacturing equipment in a manufacturing facility. This equipment processed many different parts. When the employees changed over from one part to another, it took them nine hours to change the equipment. This means that the equipment sat for nine hours not making any production while it was being changed over. I utilized a team in a collaborative approach and set a goal of reducing the set-up time by ninety percent. This means that the goal was to go from nine hours setup time to less than one hour. This set-up reduction was conducted through a group exercise in which setup time was broken down into categories. The two categories were internal and external. Internal items are items that could only be set-up while the machine was not being operated and external items were items that could be done while the machine was operating. The idea was to convert the internal set-up items to external and then reduce the external time. The team collaborated and reduced the set-up time to thirty minutes. This is a major reduction in set-up time. Prior to the set-up reduction, it took nine hours to set-up the equipment as outlined above. The department had eighteen of these machines so

## 11- DRIVE TO ACHIEVE

this was a major productivity and cost improvement for the company. This set-up reduction allowed us to reduce to six machines instead of eighteen. This was also a major reduction in cost for the department. The department costs were reduced by forty-five percent.

When you collaborate with others, make sure that you place focus on achieving a measurable outcome. Long-term success will require collaboration with others. This should be the case whether you are building a strong marriage or working for an organization. When there is a goal and measurement for the team effort, everyone involved understands their role in meeting the goal and measurement. The focused goal and measurable outcome eliminates conflicts and disagreement. Having goals and measures promote personal accountability. Accountability promotes ownership for responsibility of any mistakes that may occur. As you and your spouse or teammates work through goals and measures, make sure that you have some flexibility associated with how you proceed. Flexibility will allow the collaborative efforts to be more effective.

# HOW TO BEAT THE ODDS-BE DRIVEN!

## *Build a Stable Environment so that Everyone Can be Successful*

Stability means that the environment is firmly established and not changing or fluctuating. As an individual, it is necessary to build an environment around you that is firmly established so that you have a chance to be successful. As a college student, you must create an environment that allows you to complete your homework and study for exams. You should not plan to attend a party every night if you want to be successful with your grades because that would leave little time for you to complete homework and study for exams. You must understand the environment that is required to give you a chance to be successful.

A lack of stability prevents people from making improvements in their systems. It is necessary to have this stability in place so that you know if the environment is causing or preventing a level of success. Once the environment is stable, you can then create a standard you use so that you can repeat the results over and over again. A stable environment in your personal life can also make it easier for you to achieve a high level of success. It takes a commitment to build a stable environment.

## 11- DRIVE TO ACHIEVE

The foundational level of stability is consistency. Practicing consistency can help you build a stable environment. It helps reduce chaos in your life. There are basic items in your life that you can start with to build stability. For example, I make sure that I eat three times per day and I eat those meals at the same time each day. Another area where I build consistency into my life is with an exercise program. I make sure that I exercise thirty to forty-five minutes a day at least six days per week. I typically take one day off as a recovery day. These are two basic areas, but they help me reduce stress and allow me to control my weight. They also help me keep a clear mind and allow me to focus on tasks at hand.

## *Love and Care for Others*

It is important to love and care for others. As I watch successful people, one key ingredient I always notice is how caring they are for others. Many successful people have foundations that focus on helping people that may be less fortunate or that may need a confidence lift because they fell on hard times. This approach will help us have a better world for the next generation. The person that you help may be the next doctor who develops a world-altering cure for a disease. We should encourage others to dream big and show them the importance of believing in themselves. Make sure you are intentional when being kind to others regardless of the situation.

# HOW TO BEAT THE ODDS-BE DRIVEN!

Teach others that all situations should be used as an opportunity to be successful or learn something new. Love your family, friends, neighbors, and all humans. The Bible speaks of four types of love. They are eros, storge, philia, and agape. Eros is sensual or romantic love. This is the type of love that I have for my spouse. Storge is family love. This is the type of love that we show our children and grandchildren. Philia is love for fellow humans. It includes care, respect, and compassion for people in need. It is also brotherly love. Agape is the highest of the four types of love in the Bible. This is God's immeasurable and incomparable love for humankind. This form of love is pure, perfect, unconditional, and sacrificial. Practicing these forms of love as appropriate will help you be a better person and will help others become better stewards for our world.

Love is necessary to help build a successful atmosphere around you. Dr. King said, "Darkness cannot drive out darkness, only light can do that. Hate cannot drive out hate, only love can do that." When you get in tough situations and it is clear others may be displaying hate, do not give them hate in return, but show them that love and you will be successful with getting your message and point across. This is necessary so that you can be successful. People that show bitterness will be considered a non-team player, so never show bitterness, but display love. Dr. King also said, "Love is the only force capable of transforming an enemy into a friend." Love others

## 11- DRIVE TO ACHIEVE

and you will win them over. Your ultimate goal is to be successful. It is important not to hate because hate can lead to destruction and you want to be successful.

I try to show love to others in many different ways. I want people in general to understand that they are loved. I randomly select people in the grocery stores or restaurants to pay their bills. In a restaurant, this can be done anonymously. You never know what kind of day that someone may be having, but by paying their grocery or restaurant bill, you can show them they are loved. I also work to encourage others in the activities they are participating in, whether it is an adult on a job or a youth playing sports. It means a lot to people to know that they are loved. Life is too hard for people to have to do it alone. I also feel a sense of accomplishment when I help others. Do not be afraid to help and encourage others. When you are truly successful, you will pay it forward and help others become successful as well. Be intentional to show others love and encouragement.

## *Continuously Use These Principles*

Plan to use the twenty points I introduced at the start of this book to build a successful life and beat the odds that may be stacked against you. These principles can also help you become debt free. These items should be part of your standard mode of operation. You will

## HOW TO BEAT THE ODDS-BE DRIVEN!

experience many challenges along the journey to success. Plan to navigate these challenges and learn from them, but do not allow these challenges to get you off course. The journey is long and never ending so stay the course throughout your path. You only live once, so enjoy the journey because it will define you as a person and define your level of happiness and purpose. Remember on your journey that you want to always try to perform better than you previously performed. Successful people do not obtain success easily. Success comes after many failures. If you fail, then you need to get back up and try again. Make sure you are aiming toward success in what you want to do in life and not something others want you to do. You are the one that must be happy with the outcome of your journey.

## *Repeat the Steps*

I have found that processes work effectively when they are well-designed and robust. This book has identified a process that consists of twenty steps that are well designed and robust and they can help you on your journey towards success. It is critical that you understand that this is a journey and the journey never ends. On this never-ending journey, you need to continuously repeat these twenty steps so that you obtain more success on a continuous basis. You must refine each step continually so that you remain focused toward success. Here are the steps:

## 11- DRIVE TO ACHIEVE

1. Make God your foundation.
2. Develop strong family values.
3. Build a strong circle of friends.
4. Utilize education to help you grow.
5. Develop and live by a budget.
6. Overcome obstacles.
7. Learn from coaches and teachers.
8. Set clear goals and measure progress.
9. Drive to achieve.
10. Build trust.
11. Be visible to your family, friends, and co-workers.
12. Build quality character, friendships, products, and services.
13. Provide resources to help others be successful.
14. Coach and mentor others.
15. Work with others to build success (teamwork).
16. Be involved with the process.
17. Build collaborative relationships.
18. Build a stable environment so that everyone can be successful.
19. Love and care for others.
20. Repeat these items.

# HOW TO BEAT THE ODDS-BE DRIVEN!

## *Some Final Thoughts*

As you can tell from reading this book, having purpose is important to me; I strive to build a life that creates opportunities for others and helps them become successful. As a young boy, I grew up with many aspirations. I dreamed that I would grow up to be like Dr. Martin Luther King Jr., MacGyver, Superman, Richie Rich, and Gayle Sayers. These were all figures that I saw as relevant to the level of success that I wanted to achieve. I wanted to emulate the charismatic approach that Dr. King possessed and how he was able to launch a movement that made a significant impact on Civil Rights for black Americans. He had the leadership qualities that I would strive to gain as my approach. The non-violent approach has a power that the aggressor does not. Be mindful if you fight with an aggressor, he will try to turn things around to make it look like you are the aggressor. But if you lead in a non-violent way, the aggressor does not know how to proceed because if he becomes forceful, society will see that the aggressor is acting hateful. By remaining non-violent, all the aggressor can do is have a temper tantrum and everyone can see the childish acts that the aggressor may perform. The aggressor does not know how to handle the non-violent approach.

## 11- DRIVE TO ACHIEVE

Dr. King was a true leader and his deeds continue to help our society. Human rights must be for all humans. Our society must eliminate racist acts. All humans have the right to live a life with the opportunity of having a fair chance to be successful. When racism gets involved, human rights are oppressed and that is not fair to our society. As you read this information, make sure that you think about how you would feel if you knew that someone denied you a job because of your color. Please think about how you would feel if you were targeted and followed in a store because of your race even if you were a multi-millionaire. Please imagine how you would feel if you were being profiled by the police because of your race even if you have never been in trouble or committed any unlawful acts. Please imagine how you would feel if people did not want to include you in their groups because of your race. We must wake up and give everyone human rights. Every human deserves human rights.

I know the television character MacGyver is not a real person, but I wanted to have the skills that he displayed in the TV show. I wanted to be able to survive in the wilderness if I was dropped off deep in the forest. I wanted to be able to survive and then thrive. This is the mechanical side of me coming out. MacGyver was able to develop what was necessary to get out of situations that appeared to be doom and gloom. In the real world this is being able to get out of the wilderness. Being resourceful has always been som-

## HOW TO BEAT THE ODDS-BE DRIVEN!

ething I have strived to be so that I could thrive on my drive towards success. I also wanted the ability of another TV character, Superman. In my drive towards success, I wanted to make sure that I was able to help and save others so that they could become successful. Our world is big enough for us to help others become successful. This quality is critical to me because it allows us to build a culture for success that will spread throughout our society. The last TV character I wanted to emulate was Richie Rich. I wanted to be able to have the resources required to live a life of happiness and joy.

The last person who I wanted to emulate is a real person just like Dr. King and his name is Gayle Sayers. Gayle Sayers played football. He was an outstanding football player at all levels. He played college football at the University of Kansas and displayed great talent. He was known as the Kansas Comet. I wanted to have the level of success that he enjoyed as a college football player. He played in the NFL for the Chicago Bears from 1965 to 1971. Due to my age, I only could remember him in 1970 and 1971 as I began to play football. I researched him as I was growing up and learned he displayed leadership qualities that I wanted to have on my journey toward success. He was a hard worker and he refused to give up even after sustaining a knee injury. He displayed other characteristics that

## 11- DRIVE TO ACHIEVE

I wanted to make sure I had as a football player and through life. After football, he became an athletic director at a university. This showed me that a football player could be an effective leader in the business world after a career in the NFL.

I hope you use the information in this book to help you on your journey towards success. I am available to work with you one-on-one or in a group setting if you would like me to speak to you in person. My contact information is located in the back of the book. God is the creator and he wants us all to be successful. If you have not started your journey towards success, it is now time for you to begin. I am rooting for you.

**HOW TO BEAT THE ODDS-BE DRIVEN!**

---

*Storms will come, but you must be well anchored so the storms do not blow you away. The water from these storms should be used to help make you grow and thrive.*

*- Eddie Maddox, Ph.D.*

---

## ABOUT THE AUTHOR

Eddie Maddox, Ph.D. is an author, engineer, and professor. He has an undergraduate degree in Manufacturing Engineering from Western Carolina University, a Master of Business Administration from Southern Wesleyan University, and a Doctorate in Organization and Management from Capella University. Dr. Maddox has worked in the manufacturing sector for over thirty-four years in many key positions. These positions are as follows: Manufacturing Engineer, Product Engineer, Application Engineer, Manufacturing Supervisor, Engineering Manager, Quality Manager, and Operations Manager. He has held these positions at Ingersoll Rand (Torrington) and Timken. He also has taught on the college level at many universities. Dr. Maddox's first book *How to Beat the Odds: Be Driven!* shares lifelong wisdom for purpose-driven individuals who want to rise above the odds placed against them. He has a higher educational services company that offers the following services:

# HOW TO BEAT THE ODDS-BE DRIVEN!

- Dissertation Coaching
- Math Tutoring For All Levels
- Mentoring for College Level Classwork
- Problem Solving Coaching
- One to five week classes offered to help improve skill sets for job or educational needs for college in the following subjects:
    - Operations Management
    - Supply Chain Management
    - Human Resource Management
    - Corporate Social Responsibility
    - Strategy and Policy
    - Organizational Structure and Design
    - Quality Management

Dr. Maddox has received many awards and honors. He was an All-State High School football player. He was an Academic All-American first team football player on the college level. He also was a Magna Cum Laude graduate. He is married with two children.

Dr. Maddox is involved with events through the church, non-profit community organizations, and local school activities that are geared towards helping people learn the tools that are required to be successful.

## CONNECT

If you enjoyed this book, please leave a review on Amazon.com and Barnesandnoble.com.

## CONNECT WITH AUTHOR

### WEBSITE: MADDOXHIGHEREDSERVICES.COM

### E-MAIL: EMADDOX14@AOL.COM

### PHONE: 864-934-6993

www.ingramcontent.com/pod-product-compliance
Lightning Source LLC
Chambersburg PA
CBHW071455080526
44587CB00014B/2111